Psychedelics

Sacred Plant Medicines Healing &
Psychedelic Experiences

*(Powerful Medicines for Anxiety Depression
Addiction Ptsd and Expanding Consciousness)*

Dennis Hatch

Published By **Jordan Levy**

Dennis Hatch

Psychedelics: Sacred Plant Medicines Healing & Psychedelic Experiences (Powerful Medicines for Anxiety Depression Addiction Ptsd and Expanding Consciousness)

ISBN   978-1-998038-20-6

No part of this guidebook shall be reproduced in any form without permission in writing from the publisher except in the case of brief quotations embodied in critical articles or reviews.

Legal & Disclaimer

Table Of Contents

Chapter 1: What Are Psychedelics?

There are such a number of excellent forms of tablets obtainable that the time period "drug" has all but misplaced its meaning. While maximum human beings would love to companion drug use with healing makes use of, the reality hasn't usually been that way. There have continually been those who use drugs to benefit outcomes that can be defined as "beyond medical." Some take capsules because of the fact they experience its effects definitely enhance them in a unmarried way or some distinctive.

Psychedelic pills are defined as any kind of substance that motives an alteration in cognition and belief. These substances, lumped collectively with precise substances, together with dissociatives and deliriants, are considered to be beneath a larger institution of medication known as "hallucinogens." Once in whole effect, someone intoxicated via manner of psychedelics profits an enjoy this is qualitatively special from what's predicted while they'll be in a ordinary u.S.. The excessive can be defined as a few element near what one need to expect in "altered states," collectively with trances, ecstasy, daydreams, and near-loss of lifestyles evaluations.

According to pharmacology, most psychedelic capsules are classified under 3 chemical compound groups: tryptamines, phenethylamines, and lysergamides. Among the most common forms of psychedelics are Ayahusca, LSD (moreover called acid), DMT, Entheogens, and Peyote. These subtypes may be further broken down as we bypass alongside in this ebook.

The presence of psychedelics in society has been well documented, and we preserve to discover more societies in which they were well-known, specially as technology is growing. Used in some historic and non secular rituals, further to casual use thru the layman, they have got additionally been credited with permitting human beings to create some extremely good topics, particularly within the arts.

The past century has opened new perspectives on how the ones materials labored, and it has even introduced approximately the diversification and improvement of such drugs. A terrible attitude on psychedelic use has been fostered, often because of abuse, dependancy, and different irresponsible use. However, new studies and new perspectives are really bringing the ones capsules over again into the limelight and there may be an unstoppable momentum almost about people looking to look at this exciting thing of human records and human functionality.

Chapter 2: The History Of Psychedelics

Psychedelic pills have an extended facts that extends manner once more to the time when the first documented civilizations were installation. Different forms of materials identified as psychedelics, had been found in ancient civilizations from one in each of a kind additives of the globe.

While medical explorations approximately those materials have most effective began truely on the flip of the twentieth century, psychedelics were already being used as early as 5,000 years ago, as tested by means of way of psychedelic use of surviving tribes within the far-flung locales of the arena. We don't recognize if our historical ancestors knew exactly why, or had a easy aim in getting excessive, however they sure preferred it! For those fascinated, you could studies Terence McKenna's "Stoned Ape Theory," in which he shows that the use of psychedelics need to have probably been a key to the evolutionary device of the human race. Whether this is real or not, is a few different discussion for a far longer, more thorough ebook.

The use of psychedelics specifically revolves spherical their capability to alter conscious notion. Such altered states are frequently appeared as states of transcendence, and the act of entering into such states is considered an act of empowerment and enlightenment. During historic times, vegetation (and sometimes animals and fungi) with psychedelic results had been used in the course of events, together with rituals, celebrations, and rites. The presence of these materials have become taken into consideration to be vital and borderline sacred. As such, their sources have been handled with excessive regard of their respective societies. At the identical time, a few psychedelics have been getting used for medicinal purposes.

As treasured as these drugs had been (and nonetheless are), human beings did now not then look beyond the results of these brews, potions, and mixes, as they determined their effects as in particular an act of gods and spirits. In fact, even the term "psychedelic" did now not exist once more then. It wasn't until the early twentieth century at the same time as an in-intensity inquiry on how psychedelics labored became first started out.

It all commenced in 1935 while Albert Hoffman observed LSD, arguably the maximum available amongst all psychedelic tablets, then and now. It was this discovery that would placed psychedelics inside the leading edge, for better or for worse.

The discovery of LSD has because of the reality led scientists and medical experts to take a higher have a study the viable medicinal charge of using psychedelic tablets. By the Nineteen Fifties, the ones materials had grow to be valuable machine for psychiatrists to address conditions alongside aspect schizophrenia, alcohol dependancy, and persona problems. At the equal time, clinical research have been in complete swing during the Nineteen Fifties and 1960s, with the ones studies specifically revolving round psychotherapeutic techniques.

Nevertheless, this same exposure would possibly in the end reason psychedelics getting a terrible rap in society. By the Sixties, psychedelics, specifically LSD, had end up the drug of choice for hippies, who've been now not precisely what many humans concept of as role fashions. Ultimately, the use of psychedelics had come to be one of the important symbols of rebellion and dissent. The public opinion had gotten so terrible that it prompted america authorities to show acid into an illegal substance, with use and ownership of the drug warranting criminal prices.

This surprising flip of activities brought approximately the digital cessation of scientific research on psychedelics (at the least publicly). By the early 1970s, now banned in most nations and taken into attention a victim of political backlash, psychedelic drugs had diminished into obscurity.

It wasn't until the Nineteen Nineties that psychedelics would possibly all once more emerge into the limelight. A groundbreaking take a look at of DMT through way of Rick Strassman in 1990 ought to re-ignite hobby in the lengthy-term functionality use of psychedelics for prison abilties. With companies which includes the FDA becoming more open to investigate, the floodgates had been over again opened for a scientific inquiry on psychedelic drugs. Since then, researchers have probed the capability of psychedelics as a functionality treatment for a vast sort of ailments, and thus far the consequences have validated to be very promising.

The future is bright for psychedelics and their use. New discoveries of the functionality of these substances appear to pop up a couple of instances a three hundred and sixty five days. Most importantly, however, is that the stereotype of folks that use psychedelics (similarly to marijuana) is not as awful as it turned into a half century ago. Today, there are various a fulfillment humans within the public eye who've admitted to the usage of those materials within the direction of their lifetime. The concept that you becomes a "loser" in case you spend any time using the ones device has come to be preceding. In truth, inside the case of psychedelics, many humans are starting to suppose the alternative. Additionally, due to the fact the Internet has grown to be the maximum critical behemoth that man has ever created, it's far greater tough than ever to suppress information.

With masses of years' properly really worth of lifestyles and ancient relevance, it honestly seems inevitable that psychedelic capsules will maintain to stay on, possibly even out-lasting people. So, allow's see what the technological understanding says approximately psychedelics.

Chapter 3: The Science Behind Psychedelics

Psychedelics have existed for loads of years, however it has best been at some stage in the past century that medical inquiry inside the shape, characteristic, and feasibility of those substances has been started out out. With that said, it can be said that the technology of psychedelics stays in its infancy, even though the ones tablets had been used, arguably, for as long as human civilization has existed.

As referred to in advance, psychedelics modify cognition and notion, ensuing in altered mind states which can be notably considered one of a kind from what is in any other case predicted in a everyday kingdom of recognition. How this works varies, because it depends on the kind and quantity of the drug consumed.

There are many strategies to classify psychedelic tablets. However, the appropriate way to classify them is thru the use of the form of chemical compound to which they belong. The structure of the chemical compound has a say on how they art work and the way they motive the outcomes predicted from taking psychedelics. Almost all psychedelics are labeled underneath three households of chemical substances: tryptamines, phenethylamines, and lysergamides. To get a strong draw near of the technology behind psychedelics, it is crucial to get to apprehend the ones families of compounds and the way they act as soon as within the human body.

Tryptamines

A tryptamine is a monoamine alkaloid, which has a shape this is much like that of the amino acid tryptophan. Naturally determined in some flora, animals, and fungi, tryptamine is also observed in trace portions in mammalian brains. Known to be biologically lively, it's far believed to feature as a neurotransmitter and neuromodulator. The tryptamine molecule is taken into consideration to be the spine of compounds termed as "substituted tryptamines," a family of compounds which might be direct derivatives of tryptamine. An example of a natural tryptamine is serotonin and melatonin.

Tryptamines are said to art work in strategies: as a non-selective serotonin receptor agonist and as a serotonin-norepinephrine-dopamine liberating agent.

A serotonin receptor agonist functions thru the usage of activating serotonin receptors, consequently growing the body's sensitivity to serotonin. Psychedelics especially act on the 5-HT2A receptors. A serotonin-norepinephrine-dopamine liberating agent (SNDRA) is a substance that motives the discharge of serotonin, norepinephrine, and dopamine, which can be three of the most tremendous neurotransmitters within the human body.

In short, tryptamines particularly characteristic by causing the discharge of serotonin and dopamine even as inhibiting norepinephrine release. Psychedelics classified as tryptamines consist of DMT, bufotenin, and psilocybin.

Phenethylamines

Phenethylamines are organic compounds which is probably significantly diagnosed for being stimulants and psychoactive substances. A natural monoamine alkaloid, phenethylamines are in particular derived from the amino acid phenylalanine through enzymatic decarboxylation. This substance is determined in flora and animals, further to micro organism and fungi. Just like tryptamines, phenethylamines are taken into consideration neurotransmitters and neuromodulators. Aside from psychedelics, a few phenethylamines are classified as stimulants, hypnotics, and antidepressants.

The fundamental mode of movement of phenethylamines is just like that of amphetamine, a famous enjoyment drug. It allows the discharge of neurotransmitters: norepinephrine and dopamine. Because of this effect, the ones materials are used not only for psychedelic purposes however additionally for temper regulation and weight reduction. According to investigate, in some unspecified time inside the future of physical hobby, phenethylamine production is notably extended. It is also visible as a possibly effective diagnostic biomarker for ADHD. The maximum notable psychedelic labeled as a phenethylamine is mescaline.

Lysergamides

Lysergamides are derivatives of lysergic acid - a compound that serves as a precursor for a huge own family of chemicals known as ergolines. While lysergic acid is glaringly determined in a number of vegetation, animals, and microbes, its synthetic synthesis is considered to be a turning issue inside the history of psychedelic use. The artificial manufacturing of lysergamides is in particular finished the use of natural lysergamides. The maximum outstanding instance of these chemicals is LSD, moreover popularly called acid, that is notably considered because the defining psychedelic of the 20th century.

In spite of being one of the maximum rampantly used entertainment drugs inside the global, the mode wherein lysergamides paintings continues to be now not clean. While it's acknowledged to take impact within the vital nervous system, the mechanism for the way it works remains in big element unknown. As it stands, it is extensively believed that lysergamides characteristic the identical manner as serotonin, a neurotransmitter associated with some of vital bodily abilties. These compounds appear to intervene with serotonin receptor function however the impact though remains uncertain - for now. An advanced emphasis on psychedelic research can in the end hold the crucial thing to know-how how this works.

Chapter 4: The Effects Of Psychedelics

The effects of psychedelic capsules generally tend to differ from person to individual. Some say that it is a pleasant enjoy that they might do all yet again in a heartbeat. However, others say that it is not a remarkable experience but as a substitute a deliver of undesirable emotional ache and uncomfortable emotions. Regardless of the subjective emotions about "the experience," there are particular effects to be expected while ingesting a psychedelic substance. With each quick and long term effects, psychedelic use will motive profound modifications to how your thoughts and senses carry out.

Just like with maximum enjoyment drugs, the intake of psychedelics will motive an nearly instant impact. This "excessive" is the principle cause why humans are consuming the ones substances to begin with, and you may expect this to kick in almost proper away. Between 15 to ninety minutes after management, the quick-time period results of psychedelics emerge. The excessive can last up to twelve hours, however it is predicated upon on the dosage and the individual's tolerance tiers. Here are a number of the short-term effects to be expected proper away after intake:

Hallucinations

This is considered to be the principle impact of psychedelics. A character who consumes psychedelics can assume modifications in one in each of a type senses. Most file heightened sensations which includes brighter solar shades, sharper sounds, and finer touch. Aside from distortions in what the senses seize, a person might also additionally experience subjects that are not really there. A individual's belief of time also can alternate at the same time as they are intoxicated. Time also can appear each too speedy or too slow and the person can experience totally disoriented as soon as this trouble of the excessive wears off.

Altered Bodily States

Other than changes to a person's sensory united states of america, a number of physical and physiological abilities enjoy changes as nicely. These shifts, considered thru some as a part of the excessive, can cause someone to experience drastically precise. Because there's stimulation to the worrying system, a person can revel in an stepped forward coronary heart rate, which might also then make contributions to secondary consequences, which encompass accelerated body temperature, flushing, sweating, and eye dilation. Nervous characteristic furthermore becomes altered, inflicting outcomes together with intense feelings, anxiety, changed perception, and out of control body movements.

Just like with most capsules, extended intake of psychedelics also can have extended-term outcomes on a person's body. While short-time period effects can occasionally come to be doubtlessly lethal (together with in times of overdose), the lengthy-term effects of psychedelic use can motive a few without a doubt critical fitness problems. Most of these results are considered residual via nature, however the mechanism involved remains unknown for the maximum element. Here are a number of the critical lengthy-term results related to extended use (and abuse) of psychedelics:

Drug Dependence

For maximum humans, the body develops a sturdy tolerance to psychedelics in no time. When this happens, you can want to eat more of the drug to get the excessive you desire. At the same time, it is almost quality that you can are searching for to have greater of the drug as quickly as you have were given tried it, particularly when you have a incredible "journey." On the extreme element, you generally received't get withdrawal signs and symptoms if you prevent taking the drug, which notably lowers the danger of drug dependence for almost all of folks who strive psychedelics. Nevertheless, that is but some thing to be aware about.

Psychosis

Psychedelics can brief adjust how the mind capabilities with the useful resource of manipulating neurotransmitter activity. While this brief effect is the simplest being sought by using way of manner of maximum customers of these substances, there's a threat that prolonged abuse can reason permanent damage to numerous worried device functions. A affected man or woman who has been ingesting psychedelics lengthy-time period can enlarge psychotic signs and symptoms and signs and symptoms, collectively with visible disturbances, paranoia, and mood swings. In the worst conditions, a person may moreover even boom mind harm that resembles neurologic disorders.

While most humans are hard-pressed to discover one in every of their friends or family participants who have crucial negative prolonged-time period health issues due to psychedelic drug use, that is not an excuse to behave irresponsibly and forget approximately the dangers concerned. It is critical to recognize that psychedelics are although very mysterious to even the most professional of researchers and scientists available. If you are new to psychedelics, then commonly use with caution and beneath a more skilled character's steering.

Chapter 5: The Pros And Cons Of Using Psychedelics

Like everything in existence, using psychedelics has its very very personal share of professionals and cons. Some of the experts of the use of them lead them to so compelling to apply and raise their probability of becoming very useful products for consumption thru the general public. On the opportunity hand, some of the cons of the usage of them are extra than sufficient to compel others to turn away from using honestly, not to mention advise them for prison consumption. That being said, it'd be becoming to talk approximately the experts and cons of the use of psychedelics so you can come in your private informed forestall if those materials are for you or no longer.

The Pros of Taking Psychedelics

Unlocks Creativity

This end up one of the contributing factors for the unexpected surge of psychedelic use in some unspecified time in the future of the

Nineteen Fifties and Sixties. Just approximately every professional individual, beginning from artists and musicians to politicians and insurance makers, swear thru their use of different psychedelic materials. Many obtain as genuine with that taking those substances progressed their creativity and regularly occurring normal performance. If their frame of hard work may additionally feature proof, perhaps there's truth to this declaration, as a few LSD clients are recognized to create profound works of art work. There remains no conclusive scientific proof concerning this correlation. However, it is able to likely be traced to the potential of psychedelics to heighten concerned capabilities.

Lowers Health Risks

This is one of the extra compelling blessings of eating psychedelics in place of numerous enjoyment drugs. Developing dependence on psychedelics is reasonably unusual. At the same time, withdrawal signs and symptoms and signs and signs associated with quitting also are rare. These two elements make the use of psychedelics much less volatile than the use of numerous leisure pills, particularly even as in comparison to the prescription drug addiction epidemic in our society.

Potential Medicinal Value

Aside from sporting low dangers and minimum thing results, there is potential for the usage of psychedelic materials as a means for curing numerous ailments. Since the 20 th century, studies has been completed to see if psychedelics, including LSD and DMT, can offer lengthy-term fitness blessings.

These materials was prescribed as a way for treating and/or handling severa psychiatric troubles. Now, with the terrible stigma related to psychedelics nearly gone, scientific and clinical research on psychedelics have all yet again started out. This way that the medicinal price of those materials can nice increase in the close to destiny.

The Cons of Taking Psychedelics

Potential For Abuse

While the capability for growing dependence can be very low, there are some individuals who nevertheless turn out to be becoming dependent on psychedelics, albeit in a very specific manner. If the "trips" the person is receiving are in elegant accurate, then there is typically the functionality that they'll resort to ingesting this drug as often as viable. Good trips, visible as a reward for consuming psychedelics, can be very addictive for a few people, specifically if they may be mentally volatile and remoted from the general public.

Effects Are Mostly Subjective

Subjective outcomes are what makes psychedelic drugs a number of the greater curious materials obtainable. Each adventure has a bent to be precise and there's no telling how an person will react to the intake of a psychedelic substance. Some people revel in that the revel in is fine, because it substantially improves their abilities, social competencies, and perception. Such an enjoy is referred to as as a "accurate revel in."

Others revel in that the substance reasons most vital ache and strain. There are even instances while such trips result in crucial trouble, and such an enjoy is known as as a "horrible experience." Because of this, some of customers say that the experience of taking psychedelics notably depends at the temper of the customer and the surroundings. This is why it's far endorsed to be round a greater experienced tripper your first time or to have a sober person inside the corporation if you are tripping with severa pals.

Possibility Of Poisoning

The sheer performance of a few psychedelic substances makes them a capability overdose threat. This is particularly real for beginner customers who do not understand how a super deal of the product they might tolerate. Overdosing on psychedelics can be doubtlessly risky, so special care should be taken on the equal time as ingesting them. At the equal time, there can be moreover the capacity for poisoning if a client consumes a sub-massive substance. Both herbal and synthetic psychedelics can purpose damage, in particular at the same time as prepared the wrong manner (by way of accident or in any other case).

While it may sound like a damaged file, searching for to figure out the whole lot on your personal is what triggers most of the terrible opinions which you hear about. It might also moreover look like commonplace sense however too many people try and estimate what the right dosage need to be, consume the psychedelic in an uneasy surroundings, or use it secretly after which wonder why they revel in paranoid, unwell, or by no means want to use those materials once more. Psychedelic capsules are simply not a few aspect to be found out thru yourself, at least not till you have have been given a few revel in below your belt.

Chapter 6: Comparing Different Psychedelic Drugs

There are various forms of psychedelic pills available within the marketplace. Each of them have a unique mode of effect and their recognition is as an opportunity affected by that. Some select out one kind of psychedelic to a few other due to reasons beginning from the power and pace of impact to its accessibility. Let's test a number of the most not unusual psychedelics used round the area.

Ayahuasca

Ayahuasca is a psychedelic brew crafted from a vine of the scientific name Banisteriopsis caapi. This brew is frequently prepared by way of manner of manner of blending Banisteriopsis with other flora that encompass psychedelic homes, which incorporates Chacruna and Chacropanga (which can be appeared to include DMT, some different shape of psychedelic).

Known as a monoamine oxidase inhibitor, the consumption of Ayahausca is idea to increase the effects of materials that comprise DMT. This is probable why the aforementioned brew aggregate have grow to be organized as opposed to brewing the ones flowers for my part.

Ayahuasca became first observed in Peru, wherein tribes residing inside the Amazon rent this brew for each recuperation and divinatory purposes. We notwithstanding the truth that don't understand how long the ones human beings have been the use of it or how they have been capable of find out it first of all, but the ones Amazonians can also want to say that spirits definitely led them to the discovery and introduction of this psychedelic brew.

Preferably fed on inside the presence of shamans, many individuals who eat Ayahuasca claim to have skilled a religious awakening - a kingdom wherein they see definitely both their cause on Earth in addition to a deep perception on how they will be at their awesome. Aside from the ones profound psychedelic outcomes, the customer commonly vomits quite a bit, this is considered as a satisfactory impact - a form of purging.

Since Ayahuasca's discovery within the 1950s, non-traditional use of the substance has been gradually growing. Its use has unfold in particular in Europe and North America. In reality, there are even church buildings that have covered Ayahuasca use as a part of their ritual in great international places, which includes the Netherlands.

At the equal time, analogues of the authentic brew had been developed specially elements of the globe. However, the vital mechanism of the brew want to be retained. A MAOI (traditionally within the form of Banisteriopsis) must be in area to maximise the effect of the DMT blanketed in the blend. Without this essential combination, the popular effect may not be attained and the revel in cannot genuinely be known as an "Ayahuasca journey."

LSD

Lysergic acid diethylamide, higher appeared by its acronym, LSD, and its colloquial call, Acid, is one of the most famous psychedelics accessible. A member of the ergoline family, it's far a psychoactive drug that reasons altered notion approaches, altered senses, and non secular reviews. As one of the most well-known leisure tablets every within the beyond and now, it gained notoriety within the Sixties because of the fact the drug of preference for those who belonged to the counter-way of life movement.

Albert Hoffman first synthesized LSD in 1938 throughout a big research application related to ergot alkaloid derivatives. He by way of the usage of twist of fate decided its psychedelic residences while he via twist of fate ingested the substance. To get similarly attitude, he later intentionally ingested LSD and decided that it possessed effective psychiatric results.

By 1947, LSD have become registered as a psychiatric drug. By the Nineteen Sixties, some of well-known figures recommended using Acid.

Currently, there can be no recognized medical or healing use for acid. However, its unlawful use as a leisure drug of choice is quite wonderful because of its psychedelic effects. Classified with the useful resource of some as an entheogen, some people use LSD as a manner to reason out-of-body reviews and advantage mystical insights. Aside from the rampant leisure use, a few one of a kind motive why LSD has grow to be categorised as an unlawful substance is because of the intense aspect results related to overdose and abuse of this drug. Still, using Acid is not considered addictive.

As a synthetic psychedelic, combining diethylamine with lysergic acid makes LSD. Substances, which incorporates phosphoryl chloride, activate the acid earlier than being mixed with diethylamine. Because the producing tool (further to the additives) is normally artificial, LSD may be produced in immoderate quantities so long as the elements and critical system are there. This is probably a number one reason stress as to why this drug is as conventional as it's far at the black marketplace.

DMT

Dimethyltryptamine, popularly identified with the resource of its acronym, DMT, is a psychedelic belonging to the tryptamine family. As a structural analogue of serotonin and melatonin, it takes have an impact on by way of attaching into serotonin receptors discovered inside the mind. DMT can be ate up in loads of strategies (ex. Ingestion, inhalation, injection). It also can be consumed on its personal or it can be used collectively with one-of-a-kind tablets, which encompass a monoamine oxidase inhibitor (as visible in Ayahausca). It is one of the greater powerful psychedelics accessible, with its consequences pretty documented over the path of its discovery.

An thrilling truth approximately DMT is that it's far a compound this is glaringly synthesized with the useful resource of the human frame. Aside from this, it may moreover be placed in as a minimum sixty particular plant life. As such, it's far defined as an "endogenous psychedelic," a primary of its kind, no lots much less. Because of its endogenous presence, it is appreciably believed that this substance is important for ordinary mind function. It is likewise taken into consideration through a few human beings to be the sector's most powerful psychedelic substance. At the equal time, it's miles taken into consideration as biochemically the smallest appeared psychedelic, with a molecular weight of 188g, it definitely is handiest barely better than that of glucose.

The impact of DMT consumption is kind of immediate even as administered the right manner. Within seconds, a person can experience their body reacting to the substance. A minute after taking the drug, he/she will be able to start having signs and symptoms and symptoms and signs associated with hallucination, a region that many customers name "DMT hyperspace." The humans spherical you may now not test it however you begin seeing stuff like intergalactic factors, time or area visiting, or maybe extraterrestrial beings. This euphoria normally subsides internal 30-60 mins, with you still feeling euphoric consequences stemming from the revel in.

DMT is presently categorised as a Schedule I drug, meaning the use and sale of this drug is unlawful except while used for clinical purposes and scientific studies. However, there may be a loophole with the use of those materials at the same time as combined with one-of-a-kind merchandise to create other psychedelic merchandise, along aspect Ayahuasca. This loophole has been utilized by some of agencies already, allowing them to possess and use DMT with out being persecuted. As it's far, the unfold of DMT abuse worldwide is extraordinarily hard to regulate.

Entheogens

An entheogen is a substance this is used for any non secular or non secular interest. Highly associated with shamanic practices and so on, those compounds are normally derived from herbal assets, at the side of flowers. These materials are usually administered sooner or later of rites of transcendence and revelation, and their results are exceedingly aligned with the feelings associated with the ones sports activities. In use for hundreds of years, entheogens do not pleasant encompass psychedelics but can also moreover encompass diverse substances along side dissociatives, depressants, and stimulants.

Entheogens have a strong presence within the information and development of enjoyment drug use and addiction. These materials are mainly ate up with the aim of venture severa rituals. Seeing the outcomes of those substances on folks that consume them, there are folks that notion it is probably a first-rate idea to utilize them beyond non secular and cultural practices.

The term entheogen become first created in 1979 through manner of a group of ethnologists, botanists, and mythology experts. The term changed into a merger of Greek phrases that type of translate into "empowered via the usage of the Gods within". This is especially rooted to the truth that maximum entheogens motive feelings of thought or enlightenment, particularly at the equal time as used beneath the context of a religious or cultural occasion. Others even use the time period as an alternative call for other substance businesses, on the aspect of psychedelics and hallucinogens.

Peyote

Peyote, additionally known as Lophophora williamsii, is a cactus pleasant regarded for having psychoactive houses. Used via Native

Americans for extra than five,000 years, it is taken into consideration to be one of the oldest gift vegetation used as a narcotic substance. Native to Mexico, it is a famous entheogen utilized in severa rituals, meditations, and transcendence practices. Nowadays, peyote is popularly used for severa features, beginning from psychotherapy to psychonautics.

The major lively element of peyote is a compound called Mescaline. A psychedelic compound belonging to the phenethylamine class of hallucinogens, Mescaline has an impact this is rather comparable with wonderful psychedelics at the side of LSD and Psilocybin. First isolated in 1897 through Arthur Heffer, it has due to the truth been synthesized and used for all styles of competencies. It first won notoriety in 1955, at the same time as for a televised check Christopher Mayhew ingested 400mg of Mescaline. While the recording wasn't verified, Mayhew later referred to as it the most interesting thing he ever did.

The use of peyote is taken into consideration as a mainstay in loads of religions of the American Indians. At the equal time, cutting-edge-day religions rose up over the past centuries, with peyote gambling a important feature in lots of its spiritual practices. While this drug is most extensive on this regard, it has furthermore come to be a famous leisure drug. In truth, many artists undergo in thoughts peyote as their drug of choice, with a number of their works even alluding to it.

In maximum international locations, the usage of peyote and Mescaline is considered unlawful. However, there are exceptions made for use in a non secular context, especially in the Americas in which it's miles taken into consideration culturally big. However, going past those parameters manner use and ownership might be deemed illegal and problem to criminal sanctions. In the United States, peyote is considered a Schedule I managed substance.

Chapter 7: The Future Of Psychedelics

There was as soon as a time on the same time as nearly all public figures shied away from speakme about psychedelics. As the drug of desire for folks who belonged to the counterculture motion, there was a horrible stigma associated with those who used the ones materials. Add this to the illegal popularity of maximum psychedelics and it similarly reinforced the lousy feelings associated with the intake of those materials. Nevertheless, the popularity of these substances is starting to show spherical. It won't have regarded in all likelihood as early as some years inside the past, but now it could be said that the destiny of psychedelics is improbably extraordinary.

There end up a time even as quite an entire lot every america within the world (in particular the united states) banned the usage of psychedelics, consisting of Acid, in massive part because of its affiliation with the counterculture movement and unlawful sports activities. However, it can be said that the political climate has changed for the purpose that then. It all started with the approval of the DMT have a look at of Rick Strassman in 1990. When the FDA approved of this take a look at, this can in the end open up the door for scientists to conduct research on psychedelics and different materials deemed to be "unlawful capsules."

Because of this revived interest in psychedelic research, a revival of sorts has begun. Old-timers who were related to the golden age of psychedelic use are getting a member of forces with new enthusiasts who are intrigued to find out extra about psychedelics, therefore beginning a revolution that without a doubt could possibly deliver the ones materials decrease again into public relevance.

The blessings of such tablets, lengthy buried below the cloud of skepticism and crook troubles, are really being rediscovered and given a miles-wanted 2nd look. Therapeutic uses of medication which include LSD, peyote,

DMT, and Psilocybin are actively being examined as of this time.

Perhaps the maximum encouraging sign of the converting mind-set regarding psychedelic use is the pass again of human checking out. While animal research are correct sufficient in maximum instances, no longer some thing beats research that makes use of human subjects. Not most effective are guidelines regarding psychedelics and human use turning into plenty a whole lot much less stringent however ability subjects also are actually inclined to come to be topics to such experiments. Such a high-quality outlook is an indication that subjects can satisfactory get higher in the future.

Chapter 8: What Is Microdosing?

The method of eating very small quantities of a psychoactive chemical, most normally LSD or magic mushrooms, is referred to as microdosing. The dose is an extended way too little for it to have any impact on how some factor is perceived.

To located it each specific way, taking a microdose will not get you immoderate.

This stands in stark assessment to the more commonplace exercise of administering a "macro dosage," which refers to a big quantity taken to set off a psychedelic nation.

It isn't viable to have hallucinations or have a non secular experience at the identical time as taking a microdose. Instead, the objective is to decorate innovative output, growth professional overall performance, heighten interest, and foster go with the flow states which might be every deeper and further continuously maintained.

Benefits of Microdosing?

The concern of take a look at known as microdosing is in its infancy at this detail.

The reality of the matter is that we genuinely do not have enough information accumulated over extended intervals so that you should make any claims that can be substantiated on the functionality blessings of utilising microdoses. The majority of the proof that we've proper now may be derived from anecdotal testimonies and research which might be primarily based mostly on surveys.

However, at this very moment, there are some of immoderate-stage research being done to investigate the quick-time period and lengthy-term advantages of taking psychedelics in microdoses. In this line of studies, pioneering agencies like MindMed are setting the tempo. Recently, taken into consideration certainly one of its research on the outcomes of microdosing LSD for ADHD has moved right away to the second phase of medical checking out.

Aside from that, the following are the primary advantages which is probably generally associated with the usage of microdosing:

☐ Higher Productivity

☐ Access To Flow States Is Made Easier.

☐ More Imagination And Curiosity.

☐ A Greater Capacity For Communication And Empathy.

☐ Increased Concentration And Focus.

☐ Higher Self-Efficacy

☐ Improved Emotional Health And Connection.

☐ More Coordination And Energy.

☐ Increased Awareness Of Oneself.

☐ May Lessen Compulsive Behaviour.

☐ May Provide Relief For Cluster Headaches.

☐Offers Protection For The Nervous System And Brain.

☐Reduced Levels Of Anxiety (Especially Existential Anxiety At The End-Of-Life Stage).

☐Better Problem-Solving And Reasonable Awareness.

The findings of a survey that changed into executed on-line and posted inside the Harm Reduction Journal quantified the benefits of taking a microdose of each psilocybin or LSD. The final take a look at contained the records of 278 people who self-noted the benefits they experienced, similarly to a rating of zero–a hundred on a scale measuring the subjective charge of the experience.

What the have a look at determined come to be as follows:

The primary advantages had been mood enhancements (mentioned through 27% of respondents), improved recognition (15% of respondents), improved creativity (thirteen%

of respondents), and more ranges of self-performance (11 percentage of responders).

The Challenges of Microdosing

Microdosing comes with some headaches to take into account as properly. Some human beings record that it makes their symptoms and signs worse, which incorporates anxiety or problem concentrating, on the identical time as others say the alternative is authentic. Microdosing can, in some people, bring about emotions of anxiety and make it extra tough to hold consciousness.

The survey referenced above blanketed questions about the most huge issues related to microdosing. Respondents most frequently stated experiencing physiological soreness (in the crucial worry), illegality, and reduce strength degrees because the most tough limitations they confronted (specifically later in the day).

According to the findings of one take a look at, microdoses of LSD altered the humans' belief of the passage of time.

Chapter 9: How To Microdose

All psychedelic compounds follow the identical steps close to the microdosing way. The quantity of the dose that is taken is the number one distinction.

1. Dose: For a microdose, how an lousy lot need to I take?

A microdose is, via its very nature, a completely low dose. The suitable dosage is decided no longer handiest with the aid of way of the chemical being administered however moreover with the resource of manner of the man or woman's weight.

In maximum instances, a microdose is considered to be less than or identical to ten percent of the traditional psychoactive dose.

The following is a list of the ordinary microdoses for several precise substances:

☐MDMA — 5–10 mg

☐Mescaline — 10–forty mg

☐LSD — 8–12 micrograms

☐Ayahuasca — Variable Dose (6 mg Equivalent)

☐Magic mushrooms — hundred mg (zero.2 grams)

☐Marijuana — 2.5–five mg THC

☐Ibogaine — Variable Dose (10%)

2. Intention: Set Intention & Goals Before You Begin.

Self-improvement is, for the most issue, the driving motivation inside the once more of the practice of microdosing. Your times can either inspire or discourage you to make use of materials like these. It is as lots as you to determine whether or not or now not that is to help you in dwelling more inside the present, enhance your average ordinary overall performance at work or college, or assist you together with your innovative sports activities.

Before you get began, it could make a worldwide of distinction on your possibilities of really achieving those upgrades if you set a few dreams or intentions for your self. It may want to now not must be very complicated if it sincerely isn't always what you need it to be. It is enough to simply make a intellectual notice of the forms of benefits you intention to achieve via the use of microdosing on your recurring.

If, but, you want to be greater worried and take a extra proactive approach for your personal development and discovery, there can be no shortage of possibilities an excellent way to accomplish that.

This workout is all approximately making development in a single's very private lifestyles. The use of microdoses is first-class a tool on the way to assist us in delving more deeply, casting a much broader internet, and engaging in more on a each day, weekly, and month-to-month basis.

In addition to this, it's miles an extremely good approach for setting up and setting up new conduct and sporting activities for your life.

Utilizing bullet journaling strategies, whether or no longer on phrase-taking software program or in a physical pocket ebook, is the best approach to preserving track of your achievements and goals even as you're microdosing.

Microdosing permits people to maximize their boom in severa taken into consideration certainly one of a kind procedures, consisting of the following:

A) Journaling & Goal Setting.

Create a listing of your goals for microdosing earlier than you even get began out. Keeping the concept inside the main fringe of your mind like this may be of splendid assist to you as you skip ahead with the manner. You are unfastened to write down down as many goals as you want, however it is endorsed which you attention on accomplishing one number one cause and or three secondary goals as an alternative.

Here are a few targets to don't forget:

☐Learn a manner to perform a little issue properly.

☐Complete your e-book or blog collection.

☐Being capable of meditate for one hour every day.

☐Complete developing that software or app.

☐Obtain a b for your check.

☐Create a modern normal or dependancy.

☐Get up at 5 a.M. Every day.

☐Write 6000 phrases each day (it's far feasible to try this).

Additionally beneficial is keeping a every day magazine to file your development. Once greater, you may make this as honest or complex as you want. Some humans desire loose-form journaling, at the identical time as others select the spark off-fashion approach.

This is first-rate from the goals you listed above. You can display your feelings and your development inside the course of your desires by using the use of retaining a every day notebook. You can include as many prompts as you want proper proper here, but I've placed it's miles terrific to maintain them short because of the fact otherwise, you may no longer fill them out each day.

As a pretty easy place to begin, you can try the use of the following questions:

☐ Put your modern-day disposition on a scale from one (not right) to ten (feeling great!)

☐ Place yourself on a scale from 1 (truely cushty) to 10 (genuinely frazzled) (high-stress)

☐Describe the way you sense at this second in time.

☐   What might also you select to perform if you may most effective pick out one trouble to do these days?

B) Create Routines & Healthy Habits.

The beginning of a ordinary of microdosing is a truely ideal opportunity to hone in on the behavior and wearing activities that artwork top notch for you.

Do you preserve telling yourself that you are going to begin getting up earlier inside the mornings? More regularly, need to I workout or meditate? Consume masses a lot less coffee, would possibly you?

Your opportunities of reaching the dreams you region can significantly enhance in case you make fine modifications on your lifestyles and domesticate conduct that will help you bypass greater effectively toward conducting the ones goals.

The following is a list of a number of the maximum conventional modifications and conduct in a single's life-style that, whilst taken into consideration, may also enhance the effectiveness of microdosing:

☐ Daily exercising — Aim for as a minimum 15 minutes of immoderate-depth interest or forty five mins of low-depth interest each day (but more is even better!).

☐   Dietary modifications — You may additionally strive a regimented diet regime much like the ketogenic weight loss program, or you may sincerely stop consuming processed and sugary foods. Practicing properly sleep hygiene is going to bed at the equal time each night time and taking steps to wind down and get ready for sleep.

☐ Meditation or yoga — Even genuinely ten mins each day can also have a massive have an impact on on pressure stages and the capability to pay hobby.

☐ Learn a brand new knowledge — The use of microdosing is a top notch tool for facilitating learning. Why no longer hire this time to growth your degree of information along the manner?

3. Timing: growing a schedule for microdosing.

Microdosing yields the first-class results while practiced frequently over a few weeks or months. When getting commenced out, there are 3 components of timing that need to be taken into consideration:

☐ When have to you take your dose? What time of the day is excellent?

☐How often within the path of the week want to you're taking a microdose?

☐How lots longer need to you hold administering the microdoses?

A) When Should I Take A Microdose During The Day?

Psychedelics, although taken in low portions, have an effect that is probably defined as barely stimulating. This implies that in case you take them too past due inside the day, they may make it tough to be able to fall asleep at the proper time.

The majority of people take their microdose first issue in the morning to keep away from this issue effect.

It is important to hold in mind that the results of most psychedelics placed on off after about six hours. Because of this, you could need to time the dose in order that it offers you with the intended advantages at a time while you stand to advantage the maximum from it. This is depending on the intentions you set up inside the segment earlier than this one.

If you need to be greater efficient at art work, for example, you ought to take the microdose definitely earlier than you begin art work for the day. This is the brilliant time to attain this. It does not make a whole lot of enjoy to take your dose at 6 inside the morning if you do not regularly start working till the past due morning hours.

If developing your creativity is one among your goals and most of your innovative artwork is finished in the overdue afternoon or early night, you have to possibly take your dose in the direction of lunchtime simply so the blessings overlap with the time of day while you conduct most of your creative paintings.

In preferred, you have to try to keep away from taking your microdose at least eight hours earlier than the time that you typically visit mattress. On the instances which you take the dose, in case you locate that you have problem falling or staying asleep, you could need to make an adjustment and take the dose even earlier in the day.

B) How many days each week must a microdose be taken?

It is not really helpful to consume microdoses each day. In the equal manner which you want to exercising often, you have to furthermore time table relaxation days.

When it entails the scheduling of microdosing, there are some awesome colleges of concept:

☐ Doctor Fadiman Dose on day 1, no dose on days and 3, then repeat.

☐ A microdosing artwork week includes 5 days followed thru days off.

☐ Weeks are alternated, beginning with a dose and ending with a damage.

Within the microdosing network, there may be a fantastic deal of debate concerning the maximum appropriate dose time table. There is neither a "proper" nor a "wrong" manner to hold it out. The only difficulty that subjects is the way it makes you revel in. Everyone is one-of-a-kind, and they'll react of their particular manner.

If you find out that taking the dose for a whole week's nicely worth of labor is too much for you, either boom the big form of days which you rest or lower the quantity that you take. These timetables are best intended to function guides; in the long run, it is as plenty as you to regulate the revel in in order that it's miles appropriate on your goals.

The timetable that became installed with the resource of Dr. James Fadiman is my non-public choice (1 day on, 2 days off). He hooked up this timetable to standardize the approach of administering microdoses to accumulate records that emerge as more dependable.

There is a amazing have an impact on that stays even after the chemical has misplaced its effectiveness. After more or less weeks of using microdoses, I've determined that I preserve to sense the identical blessings even at the instances that I do not take them.

C) How extended need to I preserve administering microdoses?

The consequences of a microdose come to be extra suggested over prolonged periods while they'll be blended with the approach of goals and the choice to perform that. When thinking about the blessings of microdosing, it's miles better to recall them as a sluggish and constant improvement over time. It takes at least weeks to create new behavior, and it may take even longer until you can flow into into effective glide states on command.

In light of this, it is important to make certain which you deliver yourself good enough rest among intervals of microdosing.

The majority of people will hold their microdosing ordinary for 6 to eight weeks, and then they'll take a holiday for round one

to two months.

Chapter 10: Microdosing Guides: Substance-Based

The overarching ideas of microdosing are steady across all substances; however, the specifics of the amount, the blessings that may be expected, the extremely good technique to control the dose, and the precautions that need to be taken whilst the use of each chemical are specific to the compound in question.

Microdosing Magic Mushrooms (Psilocybin)

The maximum not unusual chemical used by humans nowadays as a microdose is shrooms, every now and then referred to as magic mushrooms. They are easy to apply, and there was a massive amount of studies executed on the correct benefits and risks of utilizing this treatment.

☐ What it is like

☐ Increased awareness and interest

☐ Higher inventiveness

☐ Greater functionality for empathy

☐ Better mood.

How Much Magic Mushroom Should You Take in a Microdose?

Microdoses of magic mushrooms commonly variety among zero.2 and zero.Five grams in weight (2 hundred to 500 milligrams).

To start, weigh out the favored quantity of raw or dried mushrooms, or take the dose inside the form of a pill. A modest scale is all this is required an excellent manner to microdose mushrooms.

If you want to deliver your pills, you can need a scale, a tablet tool (which can be offered for very little coins and is in reality non-obligatory), and some empty gel tablets. Ginger powder, L-theanine, L-tyrosine, or L-tryptamine are in reality a number of the extra additives which is probably often covered in the microdose pills that human beings produce on their very very own.

You also can get equipped-made microdose capsules on line.

There are many super species of magic mushrooms, and each one has a completely unique quantity of the psychoactive compound psilocybin. Because of this, the amount that you have to take to make an identical microdose can also exchange. Psilocybe cubensis is with the aid of an extended manner the maximum not unusual species observed worldwide. Which need to be utilized at a dose ranging amongst 200 and 500 mg.

Different Magic Mushroom Species Microdoses:

Species Name

Psilocybin Content

Microdose Range

Psilocybe cyanescens

zero.Eighty 5%

hundred – 500 mg

Psilocybe baeocystis

zero.Eighty five%

two hundred – 500 mg

Psilocybe bohemica

1.34%

a hundred – three hundred mg

Psilocybe cubensis

zero.60 – 1%

one hundred – 500 mg

Psilocybe azurescens

1.Seventy 8%

50 – 3 hundred mg

Psilocybe semilanceata

1.28%

100 – 350 mg

Psilocybe weilii

zero.Sixty one%

2 hundred – 800 mg

Psilocybe tampanensis

zero.Sixty eight%

2 hundred – 800 mg

Amanita muscaria

zero%

Not Appropriate for Microdosing

Microdosing LSD

LSD, additionally referred to as lysergic acid diethylamide, is sincerely most of the most substantially used psychedelics within the whole globe. It modified into within the 1960s whilst public figures which incorporates Timothy Leary, Ram Dass, and the McKenna brothers, Dennis and Terrence, brought it to excellent hobby.

This thoughts-altering substance is supplied inside the form of tiny squares of blotter paper that have been dipped in liquid LSD. After being positioned beneath the tongue, the tabs permit the drugs to be absorbed via the mucus membranes and micro capillaries which is probably discovered inside the mouth.

This is the second most common chemical used for administering microdoses. It is without issues available, possesses a excessive degree of safety, and there may be a great frame of studies to assist the claims that it has useful effects.

How it Feels

☐ Increased readability of idea.

☐ Boost in energy and recognition.

☐ Frequent float-states.

☐ Lower urge for meals.

What Is a Typical LSD Microdose?

Between 8 and 12 micrograms is considered to be the standard microdose for LSD.

Depending on the excellent and energy of the acid, a single microgram of acid usually tiers from 60 to a hundred in a traditional tablet. When acid is stored improperly or for prolonged periods, its first-rate will deteriorate.

Because a microdose is ready equal to as a minimum one-8th of a tablet of acid, the maximum green method for purchasing geared up microdoses is to divide a unmarried pill into eight separate quantities.

Cut the tab into quarters with a razor blade, and then lessen each sector into smaller quarters.

Microdosing MDMA

The substance referred to as MDMA, or methylenedioxymethamphetamine, is derived from amphetamine. It is a stimulant that in fashionable affects the brain's serotonin receptors if you want to exert its results.

The consequences of MDMA are not just like the ones of any of the alternative psychedelics on this listing, no matter the fact that MDMA is technically considered to be psychedelic. Although it makes a speciality of a splendid set of serotonin receptors, it activates the identical ones as magic mushrooms and LSD do.

Even at rather high doses, which aren't advised, MDMA does not purpose psychedelic consequences; even though, the hallucinations that it does produce are considerably an awful lot plenty much less immoderate than those produced with the resource of any of the traditional psychedelics.

Some humans have stated that taking microdoses of MDMA helped them experience greater related with the human beings round them, extended their mood, and helped deal with symptoms of positioned up-worrying pressure ailment, tension, and depression. [Citation needed] However, additional research is needed to determine whether or not or not MDMA use over a extended duration is steady.

How it Feels

☐Increased bodily and intellectual strength.

☐Improved consciousness and recognition.

☐Improved temper

☐Increased functionality for empathy.

Is MDMA Safe To Microdose?

There is a lot of controversy round whether or not or no longer or no longer taking a microdose of MDMA is solid. The five-HT2B receptor is a selected type of serotonin receptor, and the coronary heart includes a number of it. This substance has an affinity for this receptor, therefore it is able to be located in huge amounts there. It is possible that coronary coronary heart valve abnormalities must increase because of activating this receptor too regularly.

This is completely theoretical at this element. Since we do no longer but have good enough proof, we are not capable of state with reality whether or not or now not or whether or not or no longer repeated, low-dose use of MDMA can cause heart valve harm.

However, due to the better chance concerned, taking a microdose of MDMA is not a few thing that is advocated. If you're however interested in taking microdoses of MDMA, it's miles important which you limit the huge kind of times you're taking it to actually or 3 doses normal with week, and which you forestall use after three or four weeks with a large amount of day without work in among doses.

Microdosing Ayahuasca

The ayahuasca vine (Banisteriopsis caapi), it's miles a deliver of DMT, and wonderful Amazonian flowers are blended to create ayahuasca (generally Psychotria Viridis or Mimosa pudica). The combination creates a remarkable psychedelic experience while blended.

Although ayahuasca isn't frequently applied in microdoses, it has the various same blessings as one-of-a-type conventional psychedelics.

How it Feels

☐Heightened sense of connection.

☐A better diploma of empathy for different humans.

☐More vision and inventiveness.

☐Colors ought to probably seem more bright.

☐Greater frequency of drift states.

How Ayahuasca Works

Serotonin receptors are the number one net net web page of movement for almost all of psychedelics. This is what ayahuasca additionally accomplishes, but it moreover has a aspect that stops the breakdown of monoamine neurotransmitters like dopamine, serotonin, and norepinephrine thru inhibiting an enzyme known as monoamine oxidase. This offers ayahuasca microdoses an delivered advantage over other psychedelics with the beneficial aid of enhancing focus, hobby, and temper.

While the MAO inhibitor in ayahuasca offers greater specialised benefits for mood and focus, the DMT part of the plant gives the same blessings as psilocybin or LSD.

Chapter 11: What Is A Typical Ayahuasca Microdose?

It may be tough to decide the pleasant ayahuasca dosage for a microdose. Different ayahuasca brews must have various potencies. The awesome of the number one additives and the ratios that the maker implemented are truly what make a distinction.

One bottle of ayahuasca can also moreover only want 2 mL of liquid to deliver an powerful microdose, while some other bottle can also want 5 or 6 mL.

The purpose is to devour the same of about 6 mg of DMT, however with out brand new mass spectrometry to observe the proper efficiency of the concoction, it is quite hard to estimate how plenty liquid you will need to carry out this dose. It's doubtful that you'll have get proper of entry to to this tool in case you're developing your ayahuasca.

The best manner to decide an powerful ayahuasca microdose is first of all a completely low dosage (plenty lower than you assume desiring) and grade by grade increase

over the years.

On the number one day, begin with 1 mL of the liquid. See the way you experience, after which little by little boom the amount thru way of one mL on every occasion until you revel in moderate adjustments in perception. The dose is truely too immoderate if you could experience the outcomes of the ayahuasca. You'll apprehend the threshold for the ayahuasca you are the use of whilst you begin to enjoy a few jump forward outcomes. Set your microdose going earlier to the final dosage you took that did not reason those side effects.

Microdosing Mescaline

Peyote, San Pedro, and Peruvian flame cacti are the various hallucinogenic species of cacti, and their energetic constituent is mescaline.

This psychedelic is less well-preferred than it previously become, however that doesn't endorse it is a lot less effective than the greater notably applied psychedelics. The fact that mescaline's raw bureaucracy (the psychotropic cacti) require a big dose to hit psychoactive thresholds is the number one motive why few humans use it. To get psychoactive doses, you need to devour many grams of raw cactus right now. This cactus has an incredibly ugly taste and texture and often motives stomach misery.

All of this is unimportant in terms of microdosing. Raw psychoactive cactus makes it pretty smooth to eat low-dose mescaline.

Inducing progressive flow states, selling profound meditation, and focusing on becoming extra conscious and determined for your each day life are all made viable with the resource of way of this substance.

Compared to maximum other psychedelics, mescaline can be very grounding and has a robust tendency to make customers revel in greater gift and "inside the advocate time."

How it Feels

☐Reduced fear and stress (some file better anxiety furthermore).

☐A more potent sense of connection and empathy.

☐More highbrow stamina and electricity.

☐Improved cognizance and interest.

☐Increased health and happiness.

What Is a Typical Mescaline Microdose?

Pure mescaline frequently is available in microdoses of 10 to forty mg. Mescaline is available in a number of workplace work, every proudly proudly owning a various quantity of the natural lively factor.

The identical dose for numerous mescaline sorts and mescaline-containing cacti is damaged down in brief beneath:

Source of Mescaline

 10 mg Equivalent Dose

 forty mg Equivalent Dose

San Pedro Cactus (Dried)

 1 gram

 2 grams

Peruvian Torch Cactus (Dried)

 3 grams

 5 grams

Pure Mescaline Freebase

10 mg

forty mg

Mescaline Hydrochloride

eleven.7 mg

forty six.Eight mg

Mescaline Sulfate

13.2 mg

fifty .Eight mg

Peyote Cactus (Dried)

zero.Eight grams

1.5 gram

Microdosing Marijuana

Despite not being psychedelic, marijuana has psychoactive effects due to an active element called THC (tetrahydrocannabinol). Endocannabinoid receptors are a set of specific receptors within the brain that THC acts thru. These receptors control some of neurological procedures, such as the transmission of pain, temper, and urge for meals.

Numerous health advantages of marijuana exist. It is used to lessen pain, help with tension, decorate sleep, and boom starvation.

The high that consists of an regular psychoactive amount of marijuana merchandise, however, is not preferred via many human beings.

It has been examined that using small portions of marijuana offers maximum of the same advantages as huge doses, however without the excessive. Additionally, research have demonstrated that low-dose THC is greater powerful at treating conditions like tension, problem sleeping, and mild ache.

Some humans prefer to take delta eight THC, a great isomer of THC that manifestly has a reduced propensity to reason worried aspect results.

Even extended-time period, low-dose THC has a defensive effect at the mouse mind, in line with animal research. It is believed that regular THC microdoses can sluggish the deterioration of neurological sicknesses like Parkinson's, Alzheimer's, and more than one sclerosis.

How it Feels

☐ Minimizes the perception of pain.

☐ Fewer nauseating sensations

☐ A feeling of relaxation and carelessness.

☐ Increased creativity and popularity.

What Is a Common Marijuana Microdose?

The ordinary amount of marijuana needed to enjoy the psychoactive results is 10 milligrams. Anything above this trouble nearly simply will reason customers to revel in feelings of intoxication. A low dose is regarded to be some element this is less than 10 mg.

When it entails microdosing, the regular variety for THC attention is amongst 2.Five and five mg.

Cannabidiol, or CBD, is a kind of cannabinoid that does not produce the equal intoxicating consequences as tetrahydrocannabinol (THC). Several studies have tested that this form of mixture produces a honest bigger healing impact.

Utilizing an fit to be eaten or oil that has most effective a small amount of THC is the excellent technique to providing a microdose of marijuana. Look for gadgets which have an same amount of THC and CBD in their makeup.

Microdosing Ibogaine

Ibogaine is an indole alkaloid that motives psychosis and is observed in severa African plant species, most extensively Tabernanthe iboga.

Similar to how ayahuasca is used to address dependancy and melancholy, this plant medicine has been applied in traditional remedy for hundreds of years and has received recognition within the twenty first century.

Ibogaine is more hard to microdose than the majority of various psychedelics. Ibogaine's performance should probably variety extensively relying at the plant utilized and the extraction method.

It's furthermore hard to pick out the pleasant dose earlier than taking ibogaine in your very non-public because of the fact everybody reacts to it in any other case.

How it Feels

Increased mental clarity

A stronger capability for reflected picture (sometimes an excessive amount of).

Feeling of empathy and connection.

What Is a Typical Ibogaine Microdose?

Ibogaine microdosing should in no way be attempted without first speaking with a informed person.

It is straightforward to overdose on ibogaine due to the reality the dose varies considerably from product to product. Ibogaine desires to be used under the steering of an skilled manual due to the fact it is able to be risky at psychoactive doses.

Having said that, the bulk of human beings discover that beginning ibogaine microdoses with 1 gram of dried root bark from the Tabernanthe iboga plant is a high-quality concept.

Chapter 12: Is Microdosing Dangerous?

Dr. James Fadiman is the maximum famous researcher inside the place of microdosing psychoactive drugs. He has spent a long time discovering the benefits, dangers, and safety of administering severa medicinal drugs in little doses.

Fadiman is positive that traditional psychedelics, collectively with psilocybin (magic mushrooms), LSD (acid), mescaline, and DMT, are pretty solid.

LSD's writer, Albert Hofmann, Ph.D., fed on little quantities of the drug for maximum of his later years. He grow to be 102 years vintage at the same time as he handed away.

Numerous extra human beings have used or nonetheless use modest doses of LSD, psilocybin, and mescaline regularly with out displaying any signs and signs and signs that might be related to the psychedelic itself.

The chance of microdosing isn't always regularly tested.

Having said that, there are some theoretical troubles with the usage of even modest quantities of some materials over an extended period, which include MDMA.

Risks Associated with Microdosing

Microdosing has a few negative outcomes as nicely. Some of these terrible results run counter to the exercise's purported "blessings."

Everyone responds in a wonderful way; however the reality that one character may additionally furthermore find out that microdosing successfully reduces their anxiety, a few different individual also can enjoy even more stressful as a end result.

Accidental macro dosing is through a long way the largest danger related to the usage of psychedelics (taking a psychoactive dose). For this motive, it is crucial to make the effort to check the proper dosage for any substance you're the usage of and initially a dose drastically lower than what you expect needing for the primary session.

You can regulate the dose to maximise the results as fast as you've got a feel of methods the chemical behaves for your frame especially.

The following are a few negative outcomes of microdosing:

☐Interruptions to sensory facts (sight, sound, & touch).

☐Alterations in temperature control (feeling heat or cold).

☐Tingling or numbness within the hands and limbs.

☐Gastrointestinal pain

☐Decreased urge for meals

☐Headaches.

☐Elevated tension

☐Impaired interest and distraction.

Lower stages of energy (specifically not unusual later on within the day).

Irritability, dissatisfaction, and melancholy

Discomfort in social conditions and problem expressing oneself.

Disorientation and confusion.

Microdosing Frequently Asked Questions

The concept of microdosing is a large one which has been present approach massive growth during the last numerous years.

Following the e-book of this piece, I've gotten quite a few responses from readers who have similarly questions. This is a group of the questions that I am asked the maximum regularly approximately this project.

If there may be something that I've forgotten to embody, experience unfastened to head away a remark beneath or use the touch form on our net internet page to get in contact with any queries that you may have.

1. Is Microdosing a Risk for a Positive Drug Test?

Yes. Failure on a drug take a look at can also cease end result from microdosing.

Several days after a dose, some chemical compounds, like THC, can however be found within the bloodstream. Although the quantities are usually too low to purpose a failed drug take a look at, it is despite the fact that possible.

Other pills, like LSD or magic mushrooms, depart the body in a day or and are surely lengthy long past. As extended as you possibly did not take the dose the day of the test, it's miles quite unbelievable that you could fail a drug check at the same time as using the ones materials. But the possibility that you may err on the take a look at thing stays there.

Avoid microdosing in case you go through random drug assessments at artwork or are frequently examined.

If you have got at least weeks' take a look at in advance than your take a look at, you need to have enough time to save you taking the microdoses and allow the body sincerely rid itself of all remnants of the drug in advance than the test.

2. What Happens If I Receive a Too High Microdose?

The maximum not unusual problem with microdosing is taking too much of a microdose.

You will then enjoy the drug's euphoric effects, which can be much like taking a lesser macrodose.

Depending on how an lousy lot you devour, you may enjoy a excessive and lose the capability to pressure or paintings.

Know your dose, understand the substance you are the use of, and start with a low dose in some unspecified time within the future of your first session. You can recognition on adjusting the dose for the quality benefits as quickly as you have got a better hold near of methods the drug capabilities.

3. Is Microdosing Legal?

The majority of psychedelic pills are forbidden.

Microdoses are most probable prohibited in which you stay.

If you live in a country in which it's miles allowed or in a rustic like Canada in which it is prison for everyone above the criminal age limit, marijuana or THC can be an exemption.

Chapter 13: How Psychedelics Are Used

Many human beings use psychedelics with the resource of smoking (and inhaling) them, consuming them, or brewing them into tea. Hallucinogens had been accomplished for millennia in lots of civilizations, and a few are becoming used nowadays in religious rites to collect ecstatic or ethereal states of interest.

In the Sixties, hallucinogens have been applied in psychotherapy, however the exercise changed into in the long run terminated, at the entire for political motives.

The use of psychedelics for experimental psychiatric treatment has lately been resurrected thanks to scientific research.

The use of psychedelics within the treatment of intellectual fitness issues inclusive of tension, depression, PTSD, and particular trauma-associated situations is making a comeback within the intellectual fitness fields of psychology and psychiatry.

The majority of people can't get their hands on controlled drugs for the reason that they may be nonetheless inside the trial stages.

Therapy, prescription medication, and meditation can all assist manipulate the signs and symptoms and signs and signs and symptoms of highbrow fitness situations, but it is essential to talk approximately all of your alternatives collectively together with your scientific health practitioner.

Types of Psychedelics

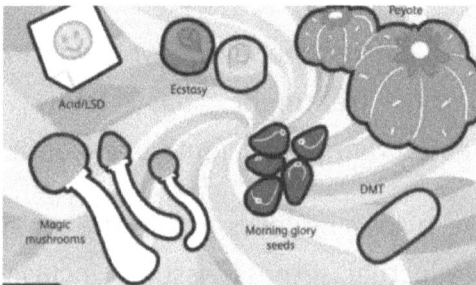

The common forms of Psychedelics are:

Acid (LSD) (LSD)

LSD, or lysergic acid diethylamide, is a synthetic hallucinogen derived from ergot, a shape of mold that grows on rye grain. LSD every now and then called "acid," cherished great reputation inside the United States until it changed into rendered illegal in the Sixties. Although LSD is now a banned chemical, its usage has persevered, albeit with ebbs and flows in popularity.

Dimethyltryptamine (DMT) (DMT)

DMT, or dimethyltryptamine, is a hallucinogenic substance determined in the bark and nuts of several Central and South American vegetation. The consequences of DMT do no longer even last up to those of various psychedelics, generally approximately an hour. Because of this, taking place a DMT ride is from time to time called a "businessman's journey" or a "businessman's lunch."

Mescaline or Peyote Cactus

It is famous that the peyote cactus, among one-of-a-kind kinds of cacti, consists of the really occurring hallucinogenic chemical mescaline. The Door of Perception," Aldous Huxley's seminal artwork on hallucinogens, statistics the research of people who took mescaline, this is similar to people who took LSD.

Although the use of peyote in valid Native American Church non secular rites is unlawful due to the drug's Schedule I category, the list does now not growth to using peyote in favored.

However, in case you make peyote for the Native American Church or distribute it on its behalf, you have to check in yearly and observe all specific crook strategies.

Ololiuqui

Plants neighborhood to Central and South America have ololiuqui, a hallucinogenic compound, of their morning glory seed pods. Though it is not a banned drug inside the United States, some humans consider ololiuqui to be a free "herbal immoderate" due to its lengthy records of use in non secular ceremonies amongst indigenous cultures in which the plant prospers.

Psilocybin

Psilocybin, the hallucinogen placed in magic mushrooms, is a obviously going on compound in numerous sorts of fungi. Because they may be decided developing glaringly in plenty of areas of the arena, hallucinogenic mushrooms are to be had a super range, and their criminal feature is murky.

The herbal origins could likely trap younger those who are keen to find out those "free materials." However, because of the doubtlessly deadly toxicity of some mushroom sorts, they offer a completely sizeable risk.

Ecstasy

Because its hallucinatory effects are an awful lot less distinguished and its temper-improving and stimulant consequences are extra obvious to the man or woman than with high-quality other psychedelics, ecstasy (or MDMA) is extra tough to find out as a psychedelic. But ecstasy can cause crazy mind and hallucinations. Even if terrible ecstasy excursions are an entire lot much less not unusual than, say, horrible LSD or mushroom trips, they do take location. Overheating, dehydration, and water intoxication are some of the situations which have been connected to ecstasy use.

Morning Glories

Morning glories are a collection of greater than one thousand flowering vegetation that belong to the equal circle of relatives as bindweeds. This circle of relatives is known as Convolvulaceae. Morning Glory vegetation are frequently used as decorations all over the international. The seeds of a number of these vegetation include d-lysergic acid amide, or LSA, that is a psychedelic tryptamine. LSA, which is likewise known as "ergine," has the same chemical structure as LSD and makes you experience the equal way.

Morning glory seeds were used as psychedelic drugs for a long term in and they may be now used in many components of the sector for the equal purpose.

Effects of Psychedelic Drugs

Psychedelic substances have extraordinary impacts on unique human beings. The effects of psychedelics variety from one individual to the following based totally completely totally on factors which includes dose, putting, and individuality.

There is a opportunity that psychedelic materials' results will encompass the following:

Alteration in temporal perception

Interpersonal conversation problems

Increased cognition or popularity Hallucinations, which includes phantom pain, noises, and/or pics

Enhanced energy

Unreasonable lack of capacity to purpose

Interactions among senses (e.G., seeing sounds)

Nausea

Having a religious awakening

stimulation of the senses

LSD, peyote, and DMT may additionally additionally moreover reason a quick rise in coronary coronary coronary heart price. A rise in center temperature might also moreover arise at the identical time as taking LSD or peyote. Dizziness, tiredness, a upward push in blood stress, a loss of urge for meals, dry mouth, perspiration, numbness, inclined thing, tremors, and impulsive behavior are all aspect consequences of LSD.

Psilocybin has a large type of outcomes, from rest to introspection to paranoia and even terror.

Peyote can cause dizziness, sweating, and flushing as factor effects. Taking DMT may additionally additionally make you feel tense and distort your sense of region and your frame.

The consequences of ololiuqui are just like the ones of LSD, no matter the truth that the substance is related to severa detrimental signs and signs, which includes dizziness, sleepiness, nausea, and vomiting.

Chapter 14: Effects Of Mood And Circumstances

Psychoactive medicinal pills have top notch results on certainly one of a type people because of factors together with the drug's motive population, the dose administered, and the individual's contemporary highbrow country (or "set") and surroundings (or "setting").

Mind, records with psychedelics, and preconceptions about the enjoy all make contributions to the "set" of a psychedelic test. For example, in case you're feeling aggravating or burdened out in advance than you're taking psychedelics, you could have a horrible time (awful revel in).

The context wherein psychedelic drug use takes location is composed of factors which embody the familiarity of the surroundings, the presence or absence of different human beings, and the presence or absence of ambient noise and artificial lighting. Using psychedelics in a peaceful, quiet, and fun setting, for example, can also reason or contribute to a terrific enjoy, whilst doing so in a chaotic, overcrowded one may have the alternative impact.

Taking psychedelics whilst in a tremendous body of thoughts, surrounded thru supportive people, and in a constant setting can assist mitigate the possibility of a terrible experience.

Bad Trips

The time period "bad enjoy" refers back to the terrifying and scary hallucinations that could occur sometimes. This would in all likelihood result in worry and erratic movements like crossing the road or maybe suicide tries.

If you operate an entire lot of psychedelics or have a in particular superb batch, you can have a few unpleasant side results.

Flashbacks

The flashback is the most traditional lasting impact of psychedelic utilization. Reliving the results of a drug, or having a "flashback," can seem severa hours, days, months, or perhaps years after the primary publicity.

The use of different materials, in addition to strain, weariness, or bodily exertion, can all motive flashbacks. Feelings associated with having a flashback might be some thing from comforting to terrifying. They regularly go through for a couple of minutes and are visible.

Mixing Psychedelics With one of a kind Drugs

Mixing psychedelics with great substances, such as alcohol, prescription prescription drugs, or possibly over-the-counter remedies, might in all likelihood have surprising effects.

Combining psychedelics with stimulants can also have a multiplicative impact on the stimulant impact, main to possibly volatile will increase in coronary heart charge and strain on the body. Anxiety is every different negative aspect impact that may be delivered on via using stimulants.

Anxiety, depression, and a racing coronary heart charge can all be exacerbated while psychedelics and benzodiazepines are used at the same time.

Reasons For Microdosing Psychedelics

The micro dosage has a exceptional goal from the recreational use of psychedelics or hallucinogens. To look at more approximately oneself, for a laugh, or for non secular or non secular reasons, many humans use psychedelics at therapeutically-advocated quantities. Conversely, microdorers generally use the ones portions to beautify their fitness or modify their mood. It's no longer till masses better doses than the effects come to be obvious like they could on a psychedelic experience. Many human beings argue that it is not the least bit like a psychedelic experience and that it's far best an imaginary emotion. Proponents of microdosing argue that consuming psychedelics in low doses produces extremely good effects on intellectual functioning.

Prehistoric Use of Psychedelic Substances

Cacti, seeds, bark, and roots of numerous vegetation and fungi have all historically been used by human beings to create hallucinogenic tablets.

It's been not unusual exercise for shamans and medicinal drug men to lease psychedelics as a portal to the afterlife because of the reality that historic times. Westerners usually have a propensity to look the work of shamans and remedy guys as normally spiritual, however many societies' entheogenic or shamanic ceremonies furthermore include additives of psychotherapy remedy.

Extensive studies into the ability chemotherapeutic and intellectual advantages of psychedelic materials had been completed with the resource of scientists in several worldwide places sooner or later of the 1950s and 1960s. Six worldwide seminars and dozens of guides had been published with regards to psychedelics by means of the mid-1960s, and by means of the usage of then there have been over one thousand peer-reviewed scientific articles outlining using psychedelic substances (given to round 40,000 patients).

Psychoanalytic strategies have been belief to be aided by psychedelic materials, making them an attractive opportunity for sufferers struggling with illnesses like alcoholism. Despite this, plenty of these research lacked the rigorous approach that is now expected.

Alcoholics, autistic kids, and people with terminal ailments are without a doubt some of the subsets of patients who have benefited from psychedelic treatment.

Regulations In The 20th century

Medical and psychiatric have a take a look at with psychedelic chemical materials changed into severely constrained throughout the

Sixties in reaction to growing troubles about the large illegal use of these drugs with the resource of the overall population (and, most famously, the counterculture).

In response to authentic issues, many nations have each banned LSD or appreciably constrained get entry to to it.

With the passage of the Controlled Substances Act in 1970, all studies within the United States into the healing capability of psychedelics got here to a halt. The United States Drug Enforcement Administration has categorized LSD and other psychedelics under the maximum stringent "Schedule I" class. Substances categorised as Schedule I are prohibited from any use inside the United States due to their excessive hazard of abuse and dependency and shortage of regular scientific use.

Authorized research into medicinal uses of psychedelic materials come to be halted globally by means of manner of way of the

Nineteen Eighties, irrespective of competition from the scientific community.

Psychedelic research and remedy schooling continued clandestinely within the a few years that located the drug's extensive ban.

Some therapists took advantage of short intervals of availability earlier than a few psychedelic substances were scheduled. Underground psychedelic treatment networks covered durations led with the useful useful resource of every trained professionals and self-taught members of the community.

Chapter 15: A Resurgence Within The First A Long Term Of The 21st Century

Clinical studies focusing on the psychopharmacological outcomes of psychedelics and their ability recuperation uses have prolonged due to the fact the early 2000s whilst there has been a resurgence of interest in their psychiatric usage. Scientific improvement has enabled researchers to build up and observe large amounts of records from animal experiments, and the improvement of imaging strategies like PET and MRI has allowed us to research wherein inside the mind hallucinogens exert their outcomes.

In addition, research at the consequences of psychedelics at the human mind has been undertaken using voluntarily collaborating drug clients as subjects, avoiding the logistical hurdles inherent in handing over illegal materials to have a take a look at individuals. As the cutting-edge millennium commenced out, there has been a famous shift in governmental thoughts-set closer to psychedelic treatment, most notably on the Food and Drug Administration.

Applications

Psilocybin (the vital lively element in magic mushrooms), lysergic acid diethylamide (LSD), and mescaline are all psychedelic capsules with ability medicinal use (the principle energetic compound in the peyote cactus).

Some research have established that those drugs are useful in treating highbrow health troubles along side PTSD, despair, alcoholism, and cluster headaches.

LSD, DMT, psilocybin, mescaline, 2C-B, 2C-I, five-MeO-DMT, AMT, ibogaine, and DOM are all examples of well-known psychedelic chemical substances which have been applied till the current-day. However, there's nevertheless loads of thriller round those drugs.

Alcoholism

A capability role for psychedelic remedy within the remedy of alcoholism (or, an awful lot less typically, exceptional addictions). The have a look at on the efficacy of psychedelic remedy for alcoholism as tons as that point changed into plagued through methodological troubles, making it now not viable to attract any organization conclusions on its usefulness. At 2–3 months and 6 months put up-treatment, this LSD treatment effect on alcohol abuse endured, however through manner of one year, it turn out to be now not statistically huge.

There changed into a statistically full-size high-quality impact of LSD at the number one stated comply with-up, which severa among 1 and 3 months following discharge from each treatment software program, some of the three studies that cautioned entire abstinence from alcohol intake.

Terminal Illnesses

Patients with terminal diagnoses can also be afflicted by excessive tension and unhappiness, however psychedelic medicinal capsules like psilocybin and LSD can help.

However, there were some minor studies finished within the 21st century due to a renewed hobby inside the use of psychedelic substances to relieve the struggling related to a terminal infection. When added in a psychotherapeutic environment and below clinical supervision, the medication is quite stable and effective for reducing degrees of anxiety and melancholy in this patient populace, furnished they have been properly screened.

Researchers within the concern of psychology who are analyzing the effects of psychedelic drug remedy on terminally sick sufferers have determined that the psychological traumatic situations of demise are frequently extra excessive than the physical ones. Because of this outlook, terminally ill people regularly battle to find out happiness and fulfillment in their very last weeks, months, or years.

Research individuals questioned all recommended elevated "clarity and self guarantee in their non-public ideals and priorities and a glowing or greater exceptional experience of fundamental that means and really well worth of life" at the same time as on the ones materials. Newer research have cautioned that psychedelic remedy should probably help the ones people face dying with a good deal less anxiety.

Cancer patients' reviews with psychedelic treatment for disappointment and tension. Both medical medical doctor and affected individual reports confirmed good sized reductions in tension and depression following psychedelic treatment, and those effects endured for as a minimum 6 months.

Given the demanding situations with powerful blinding in studies of MDMA- and psychedelic-assisted psychotherapy, it's miles in all likelihood that the outcomes are overstated.

Furthermore, no medical trials are comparing MDMA-assisted psychotherapy to modern-day proof-based totally honestly treatments for PTSD (superiority or non-inferiority), however given the results said in clinical trials of MDMA-assisted psychotherapy for PTSD, there may be no motive to don't forget that this treatment modality is more effective than gift trauma-targeted mental treatments.

Disorders of Depression and Anxiety

Ketamine intranasal for the remedy of MDD and remedy-resistant depression (TRD) in combination with an oral antidepressant turn out to be legal through the usage of the Food and Drug Administration in 2019.

The Food and Drug Administration (FDA) has unique psilocybin as a "leap forward remedy" for the treatment of critical depressive infection and remedy-resistant depression, which could probably accelerate the clearance device.

Drugs that display promise in early scientific trials and were given the "jump ahead remedy" label are given precedence for in addition research due to this expectation that they'll be drastically more a success than present day-day remedies.

Standard Psychedelic Treatment

The use of moderate to immoderate dosages of psychedelic chemical compounds is the primary approach employed inside the cutting-edge revival of studies, every now and then known as psychedelic treatment.

For the majority of the drug's first impact, patients listen to preselected tune at the same time as mendacity down and exploring their internal experiences. Conversing with the therapists is vital now not really in some unspecified time in the future of the drug session(s), but furthermore at some level in the previous education session and the afterward integration session.

In maximum times, the psychedelic enjoy is facilitated through a male and woman institution. Transcendental, mystical, or pinnacle critiques are a commonplace aspect effect of psychedelic remedy, in particular while humans take slight to immoderate dosages. Studies have demonstrated that the intensity of those sports, collectively with their speak rapidly thereafter in a treatment session, possibly a primary purpose strain of the lengthy-term consequences of signs and symptoms.

Psychological treatments consisting of cognitive behavioral remedy (CBT) and motivational enhancement treatment have been incorporated into certain psychedelic treatment studies (MET). Patients are given the danger to revel in altered cognitive and emotional states inside the context of a based CBT intervention and a dosage of psilocybin. These psychedelic effects allow for a powerful cognitive reframing of probably horrible beliefs and adjustments to at the least one's revel in of self.

Patients in a MET setting are advocated to take a look at their complicated behaviors, which includes alcohol abuse, to make modifications. Potentially, it'd moreover reduce ambivalence about behavioral modifications and increase motivation to make those modifications. This may be completed at some point of psychedelic drug evaluations with the aid of reevaluating one's revel in of self and reestablishing connections to at the least one's maximum important beliefs and standards.

Chapter 16: Psychosocial Treatment

Treatment for psychosis frequently entails taking psychedelics generally over some weeks at low to mild dosages. At the peak of the experience, the therapist is there to provide guide and help the affected man or woman art work thru any difficult emotions or mind that might occur. Patients tormented by neurotic and psychosomatic conditions were the most not unusual recipients of this sizeable category of treatment.

The assumption that this shape of remedy may also moreover resolve internal tensions offers upward thrust to the remedy's possibility moniker, "soul-dissolving" psychotherapy. Historically, psychotic remedy modified into a good sized approach of psychedelic psychotherapy in Europe, and it became additionally employed with the useful resource of using powerful psychotherapists within the United States.

To delve similarly into the subconscious, psychedelic substances might be useful due to the fact a touch a part of the adult ego generally remains functional inside the route of the experience. All the at the same time as, the affected person's thoughts is active and engaged, and the experience is one that they may consider in top notch detail later.

When human beings are in this form of reflective mode, they are acutely aware about their ego protection mechanisms, which includes projection, denial, and displacement, as they respond to themselves and their alternatives.

The final cause of treatment is to provide a together supportive surroundings in which deeply felt memories can be revisited and processed with the aid of the ideas of authentic psychotherapy. In this pretty reflective situation, the therapist can guide the affected individual in the direction of formulating a modern-day worldview or guiding philosophy for dwelling that places the onus for development squarely at the shoulders of the man or woman.

Healing Through Anaclitics

Primitive, childish cravings and impulses for a pre-genital love object are described as anaclitic (from the Ancient Greek 'anaklitos' - "for reclining"). While this form of remedy shares many similarities with psychotic treatments, it differs in that it emphasizes the ones instances in which the affected person is faced with primal sensations of emotional deprivation and dissatisfaction coming from unmet infantile demands.

To alleviate the affected man or woman's re-expert soreness, the remedy modified into designed to right away address the affected individual's suppressed need for love, bodily contact, and different primal wishes. In evaluation to the same old psychoanalyst's dispassionate stance, the therapist in this model actively participates in the remedy consultation.

Therapeutic Hypnosis

As its name shows, hypnodelic treatment have become created to increase the effectiveness of hypnotic guidelines through fusing it with the psychedelic excessive. Once the affected person had been knowledgeable to answer to hypnosis, LSD would get hold of to them, and they might be located proper proper into a trance at some degree in the drug's onset phase. There is a synergy among both techniques that makes them more powerful than each one by myself. In the 21st century, there was a resurgence of interest in the recuperation use of psychedelic capsules.

Psychedelic Tourism

Psychedelic holidays are becoming increasingly famous. While Central and South America have lengthy been associated with psychedelic tourism, cutting-edge day years have visible a increase in the affect of western way of existence on community customs. In the Netherlands, visitors can also additionally pay something from $500 to $1,two hundred to participate in a psychedelic society retreat targeted on a ritual wherein they all devour magic mushrooms and journey collectively for approximately six hours.

Underground psychedelic "publications" who declare to assist humans via their enjoy in a manner need to that of shamans in distinctive cultures are also appearing in severa factors of the USA.

HOW TO MICRODOSE PSYCHEDELICS

The reality that psychedelics may be bought legally is the primary aspect properly well worth making. Although psychedelics have come to be more and more lawful or a splendid deal a lot much less criminalized in some regions, they may be nonetheless outlawed in lots of others. To that surrender, in case you're eager to release, I'll outline the stairs you should take.

Never take day by day microdoses of any psychedelic, irrespective of how tempting. This can cause tolerance, which reduces the substance's efficacy. As a stop quit end result, most plans offer periodic breaks among microdose administrations.

Microdosing protocols evolved via manner of Fadiman and Stamets are of the maximum widely used techniques.

THE FADIMAN PROTOCOL

Microdosing is a part of the Fadiman regimen and is completed every 3 days. After the initial day of microdosing, next days can be skipped. On day four, you reduce your dosage even further. Ten rounds of the Fadiman routine take round a month to complete. After the primary month of microdosing, you ought to take a damage to evaluate whether or not you experience it.

PROTOCOL OF THE STAMETS

Paul Stamets is a mushroom professional and the writer of the Stamets protocol. Stamets is a properly-appeared expert in the place of mycology with a wealth of information. He has devoted over 30 years of his existence to coming across and growing the use of medicinal mushrooms. The 'Stamets Stack,' a manner of microdosing he created, has gained reputation. The 'Stack' is a microdosing ordinary that alternates among four days on and 3 days off of taking psilocybin (the number one chemical in magic mushrooms), Lion's Mane (a non-psychoactive medicinal fungus), and niacin (a B vitamins). The advantages of the stack components are stated in this article from the Microdosing Institute.

A Dosage Schedule Of Every Other Day

The every-one-of-a-kind-day dosing schedule is now a truth. Once every day on day one, then times every day on day , as soon as every day on day 3, and as soon as every day on day four... The listing goes on and on.

No depend a few thing approach you use, it is important which you be privy to your body and take common breaks from microdosing to evaluate the excellent effects. Most recommendations endorse taking a damage each to 4 weeks after microdosing for four to 8 weeks.

A Dosage Determined By Intuition

After following a strict manner and pausing for some time, many microdosers rely upon their gut instincts even as dosing. They do not adhere to any kind of time table and instead dose whenever they experience love it. As your focus of microdosing's effects grows, you can have an less difficult time with this.

Getting Started

Picking a dependable vendor is essential at the same time as purchasing microdosing assets.

To create its natural microdosing products, Vellum Health uses severa one-of-a-type varieties of psilocybin mushrooms. The creators, who have facts in mycology, are great approximately the strength of psychedelics to alter people's attitudes toward every distinctive and the environment.

Chapter 17: Considerations

Your initial microdose should be taken on an afternoon when you won't be doing a first-rate deal of something else. Give yourself time to take in the nuances. It's moreover smart to keep away from combining one-of-a-type psychedelics and to stick to the usage of high-quality one.

Microdosing isn't a magic bullet for losing weight. However, microdosing can aid weight loss through encouraging fantastic behavioral changes. Make sure you've got were given a robust exercise plan earlier than starting your weight reduction quest. Both a eating regimen and a microdosing time desk are included.

Expert assistance is to be had if you take shipping of as real with you want it to efficiently shed pounds. We are an professional employer of nutritionists with a focus on supporting human beings shed more pounds. We can help you in conquering your weight problems, attaining your pleasant weight, and preserving it off for proper.

Schedule a unfastened 25-minute communique with us or send us a message, and we're going to get again to you as quickly as feasible. Let's have a communication approximately your troubles just so we're able to decide out if simply considered certainly one of our Intelligent Weight Loss regimens is probably beneficial to you.

NUTRITION PLAN FOR MICRODOSING PSYCHEDELICS

All factors have variable concentrations of man or woman nutrients. Fruits and vegetables, specially while eaten whole and unprocessed or slightly processed, are an super way to decorate your health and resilience to stress.

Contrarily, processed food are often decrease in vitamins and harder for the body to digest and soak up. Consequently, if your diet is composed generally of processed meals, you can now not be obtaining sufficient of the vital nutrients you need for wholesome digestion and the involved device.

Depleting your body's dietary stores is a problem effect of the use of psychedelics like psilocybin, peyote, ayahuasca, MDMA, and LSD, particularly in case you're furthermore managing pressure and strong feelings. It's feasible to get ill inside the path of or after a psychedelic encounter in case you're bad in important nutrients.

When managing hard feelings and trauma that would floor throughout a psychedelic enjoy, it might be beneficial to fuel your frame and thoughts with a nutrient-dense healthy eating plan. In the hours leading up to your psychedelic experience, it's also realistic to abstain from something that could dissatisfied your stomach or growth your risk of tension or sadness, along with coffee and alcohol.

Here are some of the most vital vitamins and a way to acquire them:

Vitamin D

Inducing feelings of contentment and well-being, serotonin is a hormone and neurotransmitter this is probably familiar to you. Feelings of disappointment, hopelessness, anger, tension, and irritability can arise whilst serotonin production inside the thoughts is impaired.

Tryptophan, an important amino acid found in substances like oats, soy, and eggs, may be transformed into serotonin via the body. However, some help is required, and here is where vitamins D comes in. According to research, this sturdy vitamins stimulates the TPH2 gene, this is answerable for the conversion of tryptophan to serotonin.

Spending time within the sun, taking a splendid supplement, or eating food like eggs, fatty fish, and mushrooms can all increase your blood levels of food plan D.

Omega-3s

Fatty acids are vital for a enormous style of capabilities within the frame. Omega-three fatty acids are critical for masses motives, but they play a specially critical characteristic in maintaining intellectual health, keeping power ranges and immunity, and maintaining wholesome cell shape. The signs and symptoms of an omega-3-terrible diet embody weariness, bad sleep, cognitive fog, and temper changes.

Focus on consuming plenty of omega-three-wealthy food to offer your frame with the fuel it wants to maintain your intellectual fitness. While salmon and mackerel are two of the best fish alternatives, a plant-based totally completely food regimen can also offer exceptional sufficient omega-three fatty acid intake via elements like seaweed, chia seeds, walnuts, soybeans, hemp seeds, and kidney beans.

By eating extra of those gadgets earlier than taking psychedelics, you'll be capable of prevent the low temper some humans enjoy after their journey and promote the uplifted, communal sensation this is often associated with a notable experience.

Folate

Serotonin, dopamine (the neurotransmitter that enables your body revel in delight), and epinephrine are all synthesized on your mind and traumatic gadget, and they all depend upon the B nutrition folate (or adrenaline). Naturally, your psychedelic trip might be ruined if your body isn't making enough of those materials (among unique topics).

Eat your veggies, specifically broccoli, brussels sprouts, spinach, kale, and collard greens, to help your body produce those neurotransmitters. Peas, kidney beans, chickpeas, and asparagus are additionally accurate property of folate.

Magnesium

One research found that at least half of of Americans are magnesium bad, despite the fact that this mineral is crucial for the proper functioning of the neurological tool. Depression, weariness, extraordinary pulse, bronchial allergies, nausea, lack of appetite, muscular cramps, and greater have all been related to magnesium scarcity.

If you eat a eating regimen wealthy in entire meals, you may without troubles boom your consumption of magnesium with little attempt. Excellent food property encompass avocados, nuts, seeds, legumes, whole grains, and darkish leafy vegetables like spinach and kale. Dark chocolate with a excessive cacao content material fabric (seventy %) is some other proper supply of magnesium.

Prebiotics and Probiotics

Everyone appears to be discussing the importance of keeping a healthy microbiome, and for correct reason. The fitness of your digestive device immediately influences your body's capability to absorb and use the vital vitamins we have spoken approximately.

Prebiotics and probiotics are dietary supplements that could assist help healthful intestine microbiota.

Prebiotics are a shape of plant fiber used for his or her useful effects on digestive fitness and immune device feature. Prebiotic-wealthy food embody onions, garlic, leeks, lentils, barley, oats, complete wheat, jicama, asparagus, and chicory root.

Probiotics are stay microorganisms that help to pinnacle off the beneficial bacteria already gift within the intestine microbiome. Fermented food which incorporates yogurt (every dairy and nondairy), kefir, sauerkraut, kimchi, tempeh, kombucha, and miso are appropriate assets of probiotics.

If you care approximately your intestine fitness and, by extension, your ordinary fitness, you need to make it a difficulty to consume sufficient portions of prebiotics and probiotics.

The way your body responds to psychedelics is a complex dynamic, of which nutrients is simplest one element. Assume a cushty and comfortable position, and paintings on taking calm, deep breaths. To get the most out of a psychedelic experience, but, it's miles crucial to take care of yourself on all fronts, no longer certainly your physical fitness.

Chapter 18: Rehabilitation

To the attentive observer it's far smooth that psychedelics have a near relationship with the improvement of awareness and spirituality in humankind and represent one issue of our symbiosis with the plant nation. We owe our existence on Earth to a balanced biosphere, and psychedelics join us right away to the facts of vegetation. Used intelligently and with apprehend, they permit us to better apprehend the workings of the psyche and to check the recuperation of the ills that plague humanity. They aren't the solution, however instead a solution.

There prevails in our manner of life a essential misconception of psychedelics that does not do justice to their instructional capability and their beneficial functions. It has been shown that almost all of psychedelics pose clearly no hazard of dependence or physiological toxicity, no longer like pills consisting of cigarettes, alcohol, cocaine and heroin. Though this doesn't propose that they may be hazard-loose, the greatest risk stays the lack of knowledge created through using way of the outrageous terrible propaganda disseminated thru the manipulative plutocratic system, which strives to blur the distinction amongst psychedelics and difficult drugs. These are splendid families of compounds that do not have some element to do with each distinctive.

Ignorance continues humans in suffering and its antidote is Truth. This is exactly the situation of this e-book: the Truth. To rehabilitate psychedelics, I will reinstate them in a context if you want to allow a higher evaluation of the offers they offer humanity. I will proportion my personal reports and thoughts. The food of the gods helped wake up me to my actual nature – in desire she can do the equal for you.

Welcome to the Family... Of Psychedelics

The time period psychedelic is a neologism derived from the Greek (psyche: soul, and delos: seen, take area) due to this "revealer of the soul," or "that which manifests the psyche." It is widely used inside the United States to explain a circle of relatives of psychotropic tablets, commonly referred to as hallucinogens, which include LSD, magic mushrooms, mescaline, ayahuasca and others. The time period became coined in 1956 through the psychiatrist Humphry Osmond in an trade of letters with Aldous Huxley. As they sought the right call for the substances whose effects on know-how of the psyche they had been studying, Huxley, in reaction to an offer from Osmond he misunderstood, had, from ancient Greek phrases (the verb phaneroein, and the noun thymos) coined the term phanerothyme which interprets to "that which makes the soul visible, appear." Personally, I select to use the word psychedelic because it higher fits my reason than psychedelism, which appears greater indistinct, impersonal and without the presence that characterizes them for me.

Psychedelics are effective stimulants that intensify mind interest and the kingdom of recognition. They all act basically the same way pharmacologically, differing from each other in length of effect or the speed with which they act.

The time period entheogen – definitely a substance that "generates the experience of God" – regarded in 1979 as an opportunity to the time period psychedelic, which turn out to be becoming tainted with the useful resource of using its affiliation with revolutionaries, deviant agencies and with the resource of manner of the popular culture of the sixties. The contemporary-day check of entheogenic vegetation is described as ethnobotany or entheobotany.

Most psychedelics can be protected in one of the following nine groups of compounds:

1 - The LSD Family: Synthesized in 1938 for medicinal functions via the Sandoz pharmaceutical commercial enterprise corporation chemist Albert Hofmann, LSD become the principle catalyst in the lower back of the launch of the psychedelic age – its

Archetype. Lysergic acid diethylamide (or N,N-diethyllysergamide) is a lysergamides own family compound derived from extracts of rye ergot. It is a powerful psychotropic hallucinogen – very small doses are enough to motive changes in notion, mood and idea. When used to this give up, it's miles usually known as LSD, a call derived from the German Lysergesäurediethylamid.

2 - Peyote, Mescaline and San Pedro: Considered in the fifties to be the maximum effective psychedelic "door openers," peyote use in Native American ritual in all likelihood goes lower lower returned extra than 3000 years, and maintains nowadays among many Native peoples of North America and Central America. Peyote (Lophophora williamsii) and San Pedro (Trichocereus pachanoi) are cacti that include numerous alkaloids, which embody mescaline, and are ingested for their psychoactive and hallucinogenic homes.

3 - Marijuana and Hashish: The oldest documented psychedelic organization

demonstrates a synergistic effect with all the others. Delta-9-tetrahydrocannabinol, THC, has psychotropic houses and is the most commonly noted molecule in cannabis.

4 - Magic Mushrooms: The easiest to find out, they'll be gently persuasive, but powerful recognition transformers containing psilocybin and/or psilocin. Their emergence inside the past due seventies reintroduced an appreciation for psychedelics.

5 - Nutmeg and MDA: This empathic compound creates moderate hallucinations and stimulates inquiry thru its discrete psychedelic outcomes. MDA, or 3,4-methylenedioxymethamphetamine, of the phenylethylamine own family, is a psychotropic substance with stimulant and hallucinogenic houses.

6 - DMT, DET, DPT and Other Short Duration Tryptamines: They variety in depth, however because tryptamines act on serotonin receptors in the important worried device, further they encompass the maximum visually

brilliant psychedelics. The American chemist Alexander Shulgin is well-known for having studied them.

7 - Ayahuasca, Yagé and Harmaline: Derived from the Banisteriopsis vines whose bark is the number one detail of this beverage of the Amazon, the ones are telepathic healers. Traditionally fed on with the resource of using shamans of the indigenous tribes of the Amazon, this family, and Ayahusca specially, is thought for its healing houses within the context of nearby beliefs and practices.

eight - Iboga and Ibogaine: An African bush applied in initiation rites and in small dose through hunters to create lengthy-lasting peace of mind. Ibogaine, or 12-methoxyibogamine, the principle psychoactive molecule extracted from the iboga plant (Iboga tabernanthe), produces first rate hallucinations.

nine - Fly Agaric, Panther Amanita and Soma: These legendary colourful mushrooms, identical detail captivating and hellish, may additionally additionally additionally were (because the extract, Soma) at the foundation of the religious idea in Homo sapiens. They have been used for the duration of Europe and Asia in ritual and shamanic contexts. Following a completely unique schooling, ingestion end up professed to purpose states of recognition that allowed verbal exchange with the spirit international. Fly agaric includes numerous alkaloids, but its psychotropic character is especially due to muscimol, an alkaloid produced all through the drying of the fungus.

A Little History

Mankind has had a symbiotic relationship with flowers for lots of years. Ethno-mycologist R. Gordon Wasson has cautioned that the unintended ingestion of a hallucinogenic organism, possibly a mushroom, brought about the number one human spiritual enjoy and to the introduction of the thoughts of divinity and the supernatural. Humanity's dating with psychedelics and the myths surrounding them is as historic as it is ordinary. They had been described in all indigenous cultures as gods, protectors, courses, allies and teachers. From an evolutionary mindset, psychedelics have completed a position in shaping human DNA for masses of years thru distinctive feature of being covered inside the human healthy eating plan, and characteristic as a stop result made a totally precise contribution to our genetic history.

As early as 3500 B.C., frescoes depicting shamans dancing and preserving mushrooms within the presence of white livestock have been painted on rock faces within the Tassili plateau south of Algeria. Several historians have additionally positioned evidence of the usage of rye ergot or psilocybin mushrooms within the Eleusinian and Dionysian rituals of historical Greece between 1100 and four hundred B.C. Mushroom-formed stones relationship from three hundred to 500 B.C. Had been placed in Guatemala. Frescoes dating lower lower back to 300 A.D. Depicting mushrooms have been determined in Mexico, indicating the existence of psychedelic cults at that point.

In 1927, R. Gordon Wasson rediscovered the ritual use of magic mushrooms in Oaxaca in Latin America. In 1955, Wasson and Allan Richard had been the number one Americans to take part in a mushroom ritual below the supervision of Maria Sabina, a Mazatec healer. The next book of Wasson's ebook, Mushrooms, Russia, and History, in 1957, ignited public hobby in this shape of ritual.

The Swiss chemist Albert Hofmann synthesized LSD-25 in 1938, and five years later determined its psychedelic homes. He fast distributed samples to psychologists and psychiatrists for them to observe its capacity in information and treating highbrow troubles.

Since the fifties, writers and poets together with Aldous Huxley, William Burroughs, Allan Ginsberg, Carlos Castanedas, Dr. Timothy Leary and Dr. Richard Alpert added psychedelics in universities and famous life-style.

Then, in 1966, the U.S. Government, terrorized thru the effective motion added on by way of psychedelics, made maximum of them illegal. It used all the physical, monetary and political way at its disposal to unfold fear and discredit the virtues of psychedelics, irrespective of research demonstrating their splendid functionality and relative protection. This repression induced the emergence of LSD and related drugs on the black market and the subsequent years found the beginning of a freeing psychedelic movement.

In 1967, Zap Comix added us Richard Crumb, Robert Williams and Rick Griffin. Painters who have been stimulated through using visions prompted thru psychedelics encompass Robert Venosa, Vali Myers, Victor Vassarelli and Pablo Amaringo.

In 1968, a present day-day era changed into born at Woodstock to the sounds of the Grateful Dead, Janis Joplin, Jimi Hendrix,

Jefferson Airplane, Santana and plenty of others. The Beatles released the LSD-inspired Yellow Submarine and The Doors, nicely, opened the doorways of perception.

In assessment to this creative revival, really all clinical research got here to a halt. Dr. John Lilly reoriented his profession to interest his studies on dolphins and his invention, the isolation tank. Dr. Stanislav Grof, one of the most lively LSD researchers, developed Holotropic Breathwork, a method used to collect a country much like that acquired with LSD. One of the few to preserve his license to offer psychedelics changed into Dr. Alexander Shulgin, a fantastic chemist to whom we owe the rediscovery of MDMA and the advent of more than 200 new psychedelic molecules. He have end up despite the fact that obliged, however, to keep thriller the experiments he have become undertaking with a hard and fast of near buddies. When he have end up too antique to maintain his studies, he positioned them in writing collectively together with his partner Ann Shulgin of their famous books, PHIKAL and THIKAL, in which he covered recipes for the psychedelics he created.

We are experiencing a 2nd wave of psychedelic research for the purpose that early nineties. Less extravagant than the hippies of the sixties, the ones concerned on this revival are annoying no longer to lose their earnings, so they hold their studies and writings extra pragmatic and powerful. The person who initiated this second wave is Dr. Rick Strassman (DMT: The Spirit Molecule), a psychiatrist on the University of New Mexico, who acquired the vital authorization in 1990 to examine the effects of dimethyl-tryptamine (DMT) on people.

The Heffter Research Institute (HRI) and MAPS (Multidisciplinary Association for Psychedelic Studies) are non-earnings research and academic groups which help scientists benefit approval, layout and finance studies on the dangers and advantages of MDMA, psychedelics and marijuana.

Marc Emery, activist and president of the British Columbia Marijuana Party (BCMP), has lengthy beyond to first-rate lengths within the quest to have pot and psychedelics legalized. In addition to walking on the political level, his company agency released and fee variety a net net web page (www.Pot.Television) which makes available loose audio and video documents (records, conferences, workshops and precise occasions, and masses of others.). Among the various recordings available are conferences with the maximum essential stakeholders in the difficulty of entheogens. The BCMP additionally funded and helped installation the Iboga Therapy House, which makes use of ibogaine to assist remedy human beings of their addictions to materials like heroin, methadone, cocaine, crack, alcohol and methamphetamines.

Archaic Revival

This rehabilitation of psychedelics, which Terence McKenna called the "Archaic Revival," fosters the convergence of severa spheres which encompass ethnobotany, spirituality and psychotherapy, in the long run predominant inside the route of a holistic angle of the character or women.

Ethnobotany is in whole swing. In truth, the amount and remarkable of records available these days make it feasible to study and cultivate a enormous type of entheogens, and the beginner ethnobotanist can learn how to recognize and harvest without cost psychedelic vegetation which encompass morning glories, mushrooms and datura, which growth anywhere. And they can also with out issues be home-grown. Thanks to the net, entheogens can be ordered directly from their worldwide locations of beginning region and information shared on them. Sites like Erowid incorporate clean and correct facts, pointers, advice, recipes, testimonials, and so on. In addition, shops specialised in ethnobotany legally offer effective psychedelics in conjunction with San Pedro, Salvia, Iboga and the plant life critical to create Ayahuasca. As prolonged as synthetic substances remain illegal, it may be advocated to expose to ethnobotany to lower paranoia and maximize the possibilities of dwelling a first-rate and worthwhile enjoy.

Entheogens had been used as religious facilitators because of the fact earliest humanity; theories which area them at the genesis of religions and non secular traditions are gaining developing credibility. Religious traditions have used, and keep to apply, psychedelics as physical aids in non secular and sacramental practice. The Vedas, for example, the oldest scriptures on Earth, talk of Soma, an elixir containing the mushrooms Amanita muscaria or Stropharia Cubensis. In India, Patanjali consists of the "yoga of herbs containing slight" as a legitimate path to enlightenment. Several religions indigenous to the Americas use peyote, magic mushrooms, datura and morning glories. Ayahuasca is used in Peru and Brazil to speak with the gods and maintain network fitness. Rastafarians use marijuana every day.

The use of such materials for non secular ends nonetheless generates active debate in recent times. For example, for the reason that sixties an ever growing large kind of human beings are brought on Buddhism due to transcendental memories catalyzed via the use of psychedelics. Because the Buddha's teachings suggest no longer to abuse intoxicants, some practitioners bear in mind them to be proscribed. They be given as actual with that one turns into a prisoner of this kind of enjoy, and that this is not the "non-being" sought thru Buddhism and Zen. Others do not forget that intoxicants want to be tolerated however no longer be abused – it's far the center way, in the long run. It is straightforward, but, that Buddhism now has to stay in symbiosis with psychedelics. Because LSD catapults us beyond our conceptual structures and releases us from them, it circumvents our addiction of identifying ourselves with our mind and places us in a non-conceptual mode right away. Mystical states received through entheogens are so just like those obtained thru traditional strategies like meditation that it will become not possible to distinguish them.

If we take a look at Stan Grof's records, or that of people who adopt a series of severe psychedelic durations within the right region and within the proper kingdom of thoughts, what we discover is that some, no longer all, however some human beings will revel in the natural causal situation, a kingdom of Oneness, a true revel in of Samadhi, and that's what makes it properly worth it.

- Ken Wilber

Jean Gebser became the number one to word the essential shape of human recognition, which added about the development of a current branch of psychology: transpersonal psychology. It examines non-normal states of consciousness: parapsychological phenomena, trance, meditation, and paroxysmal and psychedelic opinions. Abraham Maslow, Carl G. Jung, Ken Wilber and unique therapists and thinkers enriched the problem, now considered via using many to be the fourth pillar of psychology (the alternative three being behaviorism, psychoanalysis and the humanistic approach).

The capability benefit of psychedelics inside the trouble of personal growth at the identical time as supervised via conscientious experts is straightforward. Clinical research show off that psychedelics are beneficial in relieving alcoholism and other drug addictions, post-annoying strain, despair, and obsessive-compulsive infection, in addition to beneficial in coping with relational problems and countering criminal recidivism. They are also beneficial while used along aspect terminal most cancers psychotherapy, in the stimulation of meditative states and to trigger mystical studies. Many therapists secretly used psychedelics regardless of their illegality, and after establishing their effectiveness, they could not, in top sense of right and wrong, refuse such an effective remedy to their patients, regardless of the threat of imprisonment. Those that come to mind embody Alexander and Ann Shulgin, Ralph Metzer, Timothy Leary and Richard Alpert, to call a few. An unfortunate final results of this case is that experiments and results cannot be shared publicly, depriving each the general public and professionals a quantity of treasured information.

Entheogens, whilst taken via the use of licensed human beings in the right context, in suitable doses, with a correct inner nation and with suitable goals, show great capability for assuaging pain generated at the severa existential tiers of the man or woman. They facilitate the emergence of diffused states of awareness, states which is probably in the end blanketed into the being of the person every at some degree inside the experience, and in ordinary lifestyles. By dissolving the stranglehold of the logical-rational mind on the perception of reality, entheogens engender, on one hand, the manifestation and remark of content material material arising from prepersonal stages, and on the opposite, that emanating from the transpersonal.

A number one requirement for obtaining a worthwhile revel in with entheogens is honesty. As long as one is inclined to confront and remedy all uncomfortable sensations professional sooner or later of such an encounter, the apprenticeship can be of extraordinary price. A primary reason of pain all through a psychedelic enjoy is the try through way of the difficulty to preserve a self-photo that isn't always in concord with the Self. The more one has invested within the created photograph, the extra the reluctance to exchange, consequently the more the pain is probably. The disparity may be so massive and painful that the problem may additionally revel in psychotic episodes to break out the ache. The willingness to surrender to the enjoy and permit conflicts to be resolved regularly leads to a beneficial new angle on repressed feelings, hidden values, compulsions, aspirations, and beside the point behavior. In addition, as repressed psychic fabric is launched, the inner essence of the transpersonal self is permitted to appear. This can lead to a profound and ecstatic statistics of our genuine nature, and that of the cosmos.

The psychedelic revel in tells us that indoors each mother and father is residing a self-healer. Used with care and statistics they permit us to higher recognize truth and alternate ourselves consciously.

We are sitting at a pivotal crossroad in human civilization. If our thoughts-set does not exchange drastically, our destruction is inevitable. Materialism has located us in a essential situation: pollutants, crises, suffering, and injustice gnaw at humanity like a most cancers. Through its subservient appendage, the media, the manipulative plutocratic tool magnifies those issues exponentially and reasons even extra devastation.

As cautioned in advance on this chapter, psychedelics are not the solution however as a substitute a method to this trouble. They can serve to draw hobby to the subtle states of focus and offer us a higher information of reality. Psychedelics may be useful in rediscovering a healthy spirituality without dogma or middleman, and to remedy us of many ills that distract us from the essential. They need a way of life that is deliberately conscious and in harmony with the rhythm of the universe.

To have psychedelics take their rightful location among us all once more! They are an vital part of our civilization. They are pals of our essence. Psychedelics want to all another time be preferred and venerated for their actual fee. I devote this e-book to that idea.

Chapter 19: In The Beginning

I commenced being interested in psychedelics as a more youthful character. The desire to modify my reputation took root throughout the age of sixteen, possibly way to my first female friend who had already ingested a few and mentioned it from time to time. I did no longer apprehend wherein to buy any, but, so the revel in remained inaccessible to me. I expect the primary ebook to trap my eye on the undertaking have grow to be LSD: My Problem Child thru Albert Hofmann. From that 2d on I became decided to attempt LSD, however I though did not comprehend wherein to get my palms on a few. I have come to be a more youthful introvert cartoonist with few buddies, and none of them had even attempted tablets. But I stored my ears open and positioned out that a classmate became going to take a few on the Quebec Winter Carnival. I requested him if I have to come alongside and buy some at the identical time as him because of the reality I come to be decided to ultimately quench my thirst for that candy elixir of popularity. So I found him, breathless for this extended-awaited 2d once I would possibly in the end purchase LSD. The dose end up so small, but, that I felt nearly now not some thing. It modified into only late that night time even as returning domestic that the snow started out out sounding taken into consideration considered one of a kind underfoot and I felt a peculiar lightness... However no extra. I needed to be greater aware of enjoy a barely perceptible shift in my perceptions. To be sincere, my first revel in left me hungry for extra. So I promised myself to strive once more as quickly as I need to.

In the quiet suburb in which I lived on the time, you can buy tiny drugs that have been stated to incorporate seventy five mg of LSD. I knew very little approximately psychedelics, so it in no way passed off to me to take more than two at a time. I crept along, guided thru my instinct and my interest. Subsequent experiments were higher based absolutely and normally took place in my room with my excellent friend. I painted my bedroom blue in honor of the colour of those tablets. I truely have nostalgic memories of those moments of communion as we re-listened to our favorite song, perceiving it in an entire new moderate. I in the long run had the sensation I understood companies like Pink Floyd, The Doors, and so on.

At that time we contented ourselves with one ceremony a month, which gave me time to accumulate art books, to choose the tune I desired to concentrate to, and to prepare the ceremony to make sure comfort and that I must no longer have to depart my room or communicate to my parents – who suspected now not something. We patched together 3-D glasses by using way of manner of changing lenses among red and a blue shades we wore at raves or to check images. Sometimes we went outdoor and ran throughout the residence to boom our blood float to increase the depth of our enjoy. Once we even dove within the pool. What an oceanic revel in! We were daring however our safety modified into in no way compromised; I changed into already responsible and my revelations leaned inside the route of the highbrow and creative. A satisfied teen with a ardour for comics, I did now not take the ones substances to break out my issues because I did no longer have any. Innocent, curious and open, I already sensed that there had been extra diffused and exceptional tiers of truth than that of our each day lives.

While nonetheless residing with my mother and father, I expert with small doses time and again. Being reserved and shy, I had little interest in gangs and plenty preferred my drawing desk. I even have turn out to be no longer invited to sports in which greater youthful people used capsules of all kinds and drank alcohol to extra. My solitary temperament and precocious artistic vocation saved me from falling into the abuse entice or having terrible opinions. I knew no individual capable of coaching me the deeper spiritual or shamanic components of the psychedelic enjoy. I had no concept what I modified into to discover at twenty-eight when I had my first mystical enjoy.

I moved to Montreal as soon as I have become twenty-four. Because the dealers inside the metropolis did now not encourage trust, I ate up few psychedelics until the age of twenty-seven. I did start studying compulsively at twenty-5, but, as soon as I had a revelation: I become no longer cultured. I am thankful for having undertaken this undertaking, otherwise I may additionally have observed myself empty – ignorant in a society of lack of information. Thanks to the treasured advice of my female buddy on the time – a highly informed lady, thank God! – I rolled up my sleeves and have been given all of the manner all the way down to the business enterprise of acquiring a few way of life. Because I modified into unemployed, I had all sorts of time to commit myself to this assignment I felt I had to be a extremely good artist, and mainly an entire being. I began out with literature and observed it up with philosophy, psychology, records, tune, and lots of others. Inevitably I ended up with psychedelics, which had been to permit me to corroborate the know-how and integrate it.

Timothy Leary's Flashbacks and Carlos Castaneda's The Teachings of Don Juan set fireplace to the powder. From that second on I had only one choice: LSD-25. One element introduced approximately a few different and I in the end observed some – or it located me – due to the fact in this example synchronicities are fundamental. So it modified into with the pleasure of a infant on Christmas morning that I took, at age twenty-8, the archetypal LSD on Mount Royal on a adorable sunny day. Eureka! At very last! The enjoy changed into much like the ones defined within the books I had gobbled with a ardour. I took the same of one hundred mg, therefore the experience changed into in particular seen and sensory. Being in a public region I have become no longer drawn to my internal worlds and lived instead an revel in of communion with Leary, lifestyles and nature. The colours had been exquisite, flora seemed made of satiny velvet. I have become satisfied and content material.

I then determined the Psychonaut shop, which had simply opened in Montreal. At the time, the ethnobotanicals they furnished have been strange to me, but they were criminal. So I attempted them. The human beings running there have been solicitous for my safety and well-being. I had previously smoked Salvia Divinorum, the divine sage, on 3 activities. It is a effective psychedelic to be had in severa shops in Montreal, however Salvia did no longer go away me with an super have an impact on. So, as an possibility, I offered with heightened anticipation quantities of dried San Pedro, the Peruvian Torch. It took me a few tries before finding the right dose and the proper manner to ingest this foul-tasting cactus. I first attempted boiling it and drank the gooey inexperienced tea I extracted from it. This cactus has the maximum disgusting flavor of a few element I surely have tried in my existence, or perhaps in recent times as soon as I do not forget it, I get chills. Then I touchy my approach. I floor the cactus to a fine powder and taken a touch water to make small sticky balls that I may additionally need to swallow with a little water and no longer want to taste them. My desire to enjoy them changed into stronger than my repugnance. With a quasi-superhuman will I swallowed this disgusting substance and bravely resisted the nausea that overpowered me the primary hour. I had no concept it might permit me to reach the most elegant ecstasy which modified into given me to enjoy. This have become one of the most essential moments of my existence, because of the reality with forty grams of dried San Pedro, I wakened to the spirit. At that second, Psychedelic Master modified into recalled to existence.

Chapter 20: The Intelligence Agents

Superior minds are not indulgent. Do no longer count on any gratitude from them. Help them... And mind your very non-public business enterprise. A genius is testy, touchy, regularly unstable, anarchic, discontent... However he creates.

  - André Moreau

The Intelligence Agents are wonderful beings who question the beliefs systems and conventions of their time. They question authority and suppose for themselves. These splendid instructors are boulders heaved at the walls of civilization. They radiate moderate this is too regularly missing in materialistic society. These mavericks of the primary order refuse to conform with the not unusual, they're prepared to die for the reality.

Those who resonate with my way of questioning will want to recognize the Agents of Intelligence who permeated my concept, normal it, and, in some instances have been energetic in its nascence. In this financial ruin I attention on those who had a top notch effect at the map of truth I drew for myself. Through their records, boldness, creativity and intelligence, the ones instructors guided me during my experiments with psychedelics... And interest.

The 3 maximum essential pillars supporting my idea are Timothy Leary, Ken Wilber and André Moreau. These include the sacred trinity of my private pantheon. Between them they devise a subtle stability that helps me stay sound of mind and apprehend myself. Other beings of exception who add flavor and diffused shading to my questioning encompass: John Lilly, Terence McKenna, Alexander and Ann Shulgin, Carlos Castaneda, Alex Grey and Stanislav Grof.

I will content material myself with giving my most powerful private emotions approximately every of them. This listing isn't exhaustive, severa special first rate thinkers followed me at the route of truth. I omit them to hold this financial disaster brief and relevant.

Timothy Leary

Timothy Leary (1920-1996), whose nickname changed into the Pope of LSD, changed into an American psychologist, author and activist who supported the medical use of drugs. He is the maximum famous suggest of the recovery and non secular blessings of LSD. In the sixties he invented and popularized the slogan "Turn on, music in, drop out," which have become synonymous with protest and liberation.

My call to the outer spheres of psychedelia came from Timothy Leary. His ebook Flashbacks went off like a bomb in my head. I exploded with a freeing revolutionary violence introduced on with the beneficial useful resource of the prolonged-awaited meeting with someone whose thoughts become of excessive enough pace to stimulate mine. Timothy Leary have become my religious midwife. He taught me to open the doors of perception, navigate intelligently in my psychedelic stories, be God with out complexes, now not be afraid to be myself and assume the truth... At any rate. I like his abandon, his journey, his ideas, his humor and his intelligence.

By the time I even have become twenty-8, I had take a look at most of his paintings. Tim have become an intimate buddy, a lighthouse within the spiritual darkness to which I became constrained with the aid of my loss of tangible reminiscences. It grow to be at that thing, in some unspecified time in the future of my first right psychedelic experience on LSD, that I communed with him for some hours. Tim imprinted himself on my being and I integrated him in my personality as a pal of my essence.

He changed into furthermore via my element within the direction of my spiritual awakening. I had simply study The Psychedelic Experience even as Tim appeared to me in a trance delivered on thru the use of San Pedro cactus. I have become in a country of receptivity, and there he modified into, his face superimposed at the sun, smiling at me and alluring me in the direction of the slight. I went to him and became beaten with the aid of using ecstasy. Later, I met him in the path of every different psychedelic revel in in which he congratulated me on my explorations of subtle geographical areas few humans had visited thus far. I still meet him occasionally in goals. When he involves me as an electricity being I understand the ones moments are important. I experience so blessed.

I currently determined that Tim died on May 31, my date of beginning. The second I determined about this setting synchronicity, I heard him giggle in my head and say, "I knew you'd find out sooner or later, it's miles such an apparent element." Even nowadays, this wink from Leary, and it was no longer the primary, overwhelms me with an unusual pride because of the reality a subtle link unites us in this existence.

In pre-determined on human beings, from every gene-pool, neural circuits were activated (normally with out their hobby) which might be designed to manufacture destiny realities. Future gene-swimming pools....

Those who're fortunate enough to recognize their post-human genetic caste collect a stage of amazing prescience and humorous perception. They apprehend that they're Time Travelers, sincerely on foot around in past civilizations. A maximum amusing and effective role to play. While they have got little power to change the ripples of information or the waves of evolution they're capable of surf them with growing knowledge.

Such evolutionary marketers are amazing described as OUT-CASTES. They are cast out, thrown beforehand, driven up, above and past, present day hive realities.

- Timothy Leary, The Intelligence Agents

Ken Wilber

Ken Wilber (born January 31, 1949) is an American author and spiritualist. He has written on psychology, epistemology, the records of thoughts, sociology, mysticism, ecology and evolution. His paintings tries to formulate what he refers to as an "imperative principle of cognizance," and made him one of the deans of what's called "vital concept." He based at the flip of the millennium the Integral Institute.

Ken Wilber does not advise the usage of psychedelics as such. He opts as a substitute for Buddhist exercising with an emphasis on meditation.

After my spiritual awakening, I had many questions that no person in my entourage ought to answer. That is as soon as I located transpersonal psychology, of which Wilber became a outstanding decide inside the eighties. I read his paintings with passion, which led me to the idea of essential philosophy. The overwhelming effect of psychedelics combined with any extreme spiritual workout can with out troubles uproot us if we have were given not established a practice in our lifestyles that takes into attention all tiers of being - body, thoughts, soul, spirit: that is what Wilber calls the essential exercise. So I set out to encompass into my life severa sports and carrying sports that allowed me to test more deeply with psychedelics on the equal time as maintaining my lucidity and my connection with my physical frame and social reality. As such, Wilber's map of the degrees of recognition despite the fact that serves me nowadays to better recognize my memories and make clear them.

André Moreau

André Moreau modified into born in Montreal on February 8, 1941. He received a PhD in philosophy from the Sorbonne in 1966, and conducts located up-doctoral studies in epistemology with an emphasis on sexology. He posted his first e-book in 1969 and created a philosophical system, Jovialism, based totally mostly on a monistic, immaterialist and immanentist worldview. He founded the Jovialist Movement on December 14, 1970. His entire art work includes one hundred books of which over seventy were published so far.

After six years of extensive psychedelic experimentation and animation of debate groups, I started out to move around in circles intellectually for loss of locating stimulating interlocutors. I isolated myself from my contemporaries to a volatile diploma. I needed to renew myself the least bit prices, discover my selected circle of relatives. I had tried to positioned the Psychedelic Community of Montreal, however with out a strong philosophical foundation and assist I couldn't create a tangible motion. I without a doubt sensed that I must soon meet a instructor, and all I had to do changed into wait and be attentive. Just once I turned into approximately to lose my footing, some Jovialists started to attend my group communicate. They spoke to me of André Moreau and surely certainly one of them delivered us. And so it came to bypass that an Intelligence Agent in flesh and blood manifested himself to me with an proprietor's guide and a extraordinary philosophical vocabulary that have become constant with my private ideals. André Moreau seemed like a raging hurricane in my lifestyles, confirming the adage that after the pupil is prepared, the instructor appears. Shortly afterwards, André came to me in a dream to inform me he have end up going to be my professor of philosophy and I traditional with pleasure.

Before assembly André, I had found no pertinent Francophone authors inside the area of psychedelics and focus, foremost me to dive headlong into Anglo-Saxon way of lifestyles to fill the gaps inherent in mine. So it changed into no small pride to investigate that simply twenty minutes from me lived a philosopher whose audacity and genius flow some aspect like this: "The reality is that I am the most advanced trainer the Earth has ever diagnosed. I'm no longer right proper right here to maintain what may be stored – this isn't always all that important. I am proper proper here to break the system. It should no longer marvel me to find out sooner or later that no longer a unmarried guy or ladies understood me from my start to my loss of lifestyles. We do no longer recognize lightning, we're trouble to it."

His paintings inspired an initiatory dream which brought about me to restart the Jovialist Movement, dormant for two many years. In this dream, André, surrounded with the aid of a large crowd, wore a grimy, dented football helmet. I felt his disillusionment. He took off the helmet and surpassed it to me. I observed then that it contained a crystal cranium manufactured from light. The first component I did grow to be location the helmet on my head. The crowd surrounded me then and I moved some distance from André. They located me. It was with this switch of strength that the Jovialist Movement have turn out to be surpassed to me. I will never forget about approximately this initiatory dream, and due to the fact I took it very considerably, quick afterwards I entered the kingdom of the Grand Jovialist.

Being André Moreau's pupil stimulates my strength, my enthusiasm, my intelligence... And my genius. I commit myself now to apprehend and stay Jovialism at his aspect. He is capable of assist me stay in a voluntarily conscious way. He is the lucid witness of the extraordinary immensity of my being that threatens to engulf me at instances, one of the few to surely apprehend who I am and What I Am.

The entheogens have completed their art work: they activated the subtle components of my body and led me to take take a look at of them... Now it's time to play. I am getting to know to dream consciously, often consulting André, who's a dream draw close. I query him constantly about Jovialism and I permit myself be guided at the same time as essential. Psychedelics have been my teachers for a time and that they led me to a person who can educate me to be. They offer us instinct – make ecstasy, enlightenment, and subtle and causal states of focus intelligible, however it remains for us to combine and placed them into motion. Most human beings do not have a trainer.... Me? I consciously avoids detours.

Although not the number one people to my thinking, the subsequent Intelligence Agents have been moreover, of their very personal way, vital in my development.

John Lilly

John C. Lilly (1915-2001) have become a clinical doctor, psychotherapist and writer from the American West Coast. He is terrific referred to as a pioneer within the look at of recognition, which he studied along with his very personal gear: the flotation tank, communique with dolphins and psychedelic tablets. He changed into an important member of the California counter-way of life within the late sixties and early seventies.

John Lilly helped me apprehend my widespread inner worlds and the way to discover them with intelligence and area. I stay deeply stricken by his pragmatic and obsessive person. I would really like to work with ketamine and the isolation tank at some point, like he did. With his eager intelligence and audacity, Lilly driven his studies to previously unexplored heights. He dove bravely into himself and brought lower lower back expertise and sensible records of the subtle realms of recognition and made them reachable to all.

His interactions with the ECCO (Earth Coincidence Control Office) helped me deal with similar personal memories that might have anxious me, had they no longer been made acquainted with the resource of Lilly. Some chortle at me after I speak approximately my course and extraterrestrial beings, others expect I am loopy or I that I am giggling at them. But it is healthful to have the ones styles of research whilst experimenting with one's awareness. "When given freedom from external exchanges and transactions, the remoted-constrained ego (or self or individual) has property of new records from interior."

Lilly's studies with dolphins demonstrates that this planet is already home to dwelling beings with intelligence advanced to that of people. Before looking to touch extraterrestrial civilizations we need to be privy to mammals of identical or advanced intelligence to ours which includes whales, dolphins and elephants. Their brains are huge than ours and they own a language and a way of life all their personal. Humans saturate the planet with ultrasound, unaware that they've an effect at the lives of those mammals who use the ones frequencies to speak.

Carlos Castaneda

Carlos Castaneda (1925-1998) come to be an American anthropologist acknowledged for his books chronicling his alleged testimonies below the tutelage of his mentor, a Yaqui Indian via the selection of Don Juan Matus. He changed into an anthropology pupil on the UCLA in 1960 at the same time as he met Juan Matus, an Indian claiming to be a Yaqui, and he have end up his student. His paintings catalogs his studies and instructions gleaned from this relationship, whose truth remains hotly debated.

The Teachings of Don Juan became a revelation of extremely good violence. I had in no manner even imagined how effective psychedelics critiques can be, nor the quantity of the shaman's data which offers get admission to to parallel realities. After studying this e-book I had one want: to try peyote and enjoy the separate truth. I positioned I may additionally want to legally buy San Pedro (a cousin of peyote) in Montreal. I had now not however had my first mystical reviews, I turn out to be starved for in fact the, and I couldn't stand being caught in rationality to any volume in addition. I study maximum of his books, taken aback via the usage of way of what I observed and that I had in no manner heard of the Yaqui sorcerers and their way of lifestyles.

Castaneda is a splendid storyteller who helped me apprehend the manner to stay like a warrior – generally alert, Seeing in preference to simply looking. He is privy to every element: the wind route, the hen that passes, the allies coming near. He is solitary and courageous, he is not terrified of lack of lifestyles and masters the artwork of dreaming. He is the nagual, he is privy to his power animal and is aware of the way to apply him even as essential. The warrior lives inside the present due to the fact he has erased his beyond and not issues about the destiny.

Terence McKenna

Terence McKenna (1946-2000), American writer and truth seeker, have grow to be identified for his speculations on certainly one of a kind topics, collectively with the Voynich Manuscript, the origins of the human species and the newness idea, which postulated that point and an accelerating novelty fractal wave have to culminate nicely in 2012. This concept seems to draw on a combination of hallucinogenics, Gaïaïsm and shamanism.

Terence McKenna got here to me later. I had paid little interest to him until a more youthful philosophy scholar with a eager thoughts started out attending my organisation discussions. An inveterate McKenna fan, he listened to his lectures, headphones on, on the identical time as he slept. At that component I but swore by way of way of Leary and this gave rise to fruitful discussions. Leary the psychologist and McKenna the anthropologist shape very specific faculties of mind, but I in the end allowed myself to be tempted by manner of manner of McKenna's paintings, and I do no longer remorse it. His universe is rich and his thoughts relate to topics that fascinate me: 2012, the Mayan calendar, the intimate connection indigenous peoples have with psychedelics, the food of the gods and its sacramental use.

www.ingramcontent.com/pod-product-compliance
Lightning Source LLC
Chambersburg PA
CBHW060223030426
42335CB00014B/1316